# Electromagnetic Fields and Waves

*Edited by Kim Ho Yeap and Kazuhiro Hirasawa*

Published in London, United Kingdom

IntechOpen

*Supporting open minds since 2005*

Electromagnetic Fields and Waves
http://dx.doi.org/10.5772/intechopen.77420
Edited by Kim Ho Yeap and Kazuhiro Hirasawa

Contributors
Rafael Zamorano, María G. Hernández, Verónica L. Villegas, José R. Montejo-Garai, Jorge A. Ruiz-Cruz, Jesus Rebollar, Wisnu Ari Adi, Didin Sahidin Winatapura, Ade Mulyawan, Yosef Sarwanto, Yohanes Edi Gunanto, Yana Taryana, Yunasfi Yunasfi, Mashadi Mashadi, Gennadiy Churyumov, Jinghui Qiu, Nannan Wang, Najam Siddiqi, Nasser Al Nazwani, Evgeny Bondarenko, Graham Brodie, Mohammed M. Bait-Suwailam, Kim Ho Yeap, Kazuhiro Hirasawa



First published in London, United Kingdom, 2019 by IntechOpen
IntechOpen is the global imprint of INTECHOPEN LIMITED, registered in England and Wales, registration number: 11086078, The Shard, 25th floor, 32 London Bridge Street
London, SE19SG – United Kingdom
Printed in Croatia

British Library Cataloguing-in-Publication Data
A catalogue record for this book is available from the British Library

Additional hard copies can be obtained from orders@intechopen.com

Electromagnetic Fields and Waves
Edited by Kim Ho Yeap and Kazuhiro Hirasawa
p. cm.
Print ISBN 978-1-78923-955-3
Online ISBN 978-1-78923-956-0
eBook (PDF) ISBN 978-1-83880-653-8

For EU product safety concerns:
IN TECH d.o.o., Prolaz Marije Krucifikse Kozulić 3, 51000 Rijeka, Croatia, info@intechopen.com or visit our website at intechopen.com.

# Meet the editors

Kim Ho Yeap is an associate professor at Universiti Tunku Abdul Rahman, Malaysia. He is an IEEE senior member, a professional engineer registered with the Board of Engineers, Malaysia, and a chartered engineer registered with the UK Engineering Council. He received his BEng (Hons) in Electrical and Electronics Engineering from Universiti Teknologi Petronas in 2004, his MSc in Microelectronics from Universiti Kebangsaan Malaysia in 2005, and his PhD from Universiti Tunku Abdul Rahman in 2011. In 2008 and 2015, respectively, he underwent research attachment at the University of Oxford (UK) and the Nippon Institute of Technology (Japan). He is the external examiner and external course assessor of Wawasan Open University. He is also the editor in chief of the *i-manager's Journal on Digital Signal Processing*. He has also been a guest editor for the *Journal of Applied Environmental and Biological Sciences* and *Journal of Fundamental and Applied Sciences*. In addition, he has been given the university teaching excellence award and 18 research grants. He has published more than 100 research articles (including refereed journal papers, conference proceedings, books, and book chapters), which are mostly related to electromagnetics. Among his notable publications—and those of which he is most proud—are the report on the design of the receiver optics used in the Atacama Large Millimeter/Submillimeter Array telescope and the formulations that rigorously describe wave propagation in lossy waveguides.

Kazuhiro Hirasawa is a professor (Tsukuba University), Japan (2005–present), and an international collaborative partner (Universiti Tunku Abdul Rahman), Malaysia (2013–present). He received his BEng in 1964, his MEng in 1966 (Keio University), and his PhD in 1971 (Syracuse University). He was at the University of Tsukuba (1978–2005), Tokyo University of Agriculture and Technology (2005–2007), MIT (USA) (1991), and Institute for Infocomm Research (Singapore) (2002). He was an editor of the *International Journal on Wireless and Optical Communications* (2003–2007). He has published about 200 research articles (journal papers, international conference proceedings, books, book chapters, and US government reports) in electromagnetic waves and antennas. He is an IEEE Life Fellow. He obtained the best paper award at IEEE EMC International Symposium with Dr. Morioka (2010).

# Contents

# Preface

In this book, a variety of topics related to electromagnetic fields and waves are extensively discussed. The topics encompass the physics of electromagnetic waves, their interactions with different kinds of media, and their applications and effects.

Chapter 1 gives an overview of the basic framework of electromagnetism. The topics covered in this chapter include a succinct illustration of electric and magnetic fields as a result of charges at rest and in motion, the range of frequencies of electromagnetic radiation, as well as the mathematical laws that govern electromagnetics.

Chapters 2 and 3 are related to the sources of electromagnetic waves. In Chapter 2, the generation of electromagnetic waves—microwaves, in particular—in traveling wave tubes based on Cherenkov's radiation mechanism is extensively discussed. Since many modern microwave and millimeter-wave systems, used in communications and plasma physics, require high input power, the authors of Chapter 3 describe the designs of different power combiners to amplify the magnitude of the electromagnetic signals.

Chapters 4 to 8 analyze the behavior and interactions of electromagnetic fields and waves in different types of media, i.e., in general dielectric media (Chapter 4), a rotating dielectric medium (Chapter 5), magnetic materials (Chapter 6) and metamaterials (Chapters 7 and 8).

The final chapter (Chapter 9) presents an analysis on the effects of electromagnetic radiation on living things. The authors have used chick embryos in their in vivo studies.

I would like to take this opportunity to thank Ms. Sandra Maljavac, Author Service Manager, IntechOpen.

**Kim Ho Yeap**
Associate Professor,
Department of Electronic Engineering,
Faculty of Engineering and Green Technology,
Universiti Tunku Abdul Rahman,
Malaysia

**Kazuhiro Hirasawa**
Emeritus Professor,
University of Tsukuba,
Japan

Section 1

# Introduction

Chapter 1

# Introductory Chapter: Electromagnetism

*Kim Ho Yeap and Kazuhiro Hirasawa*

## 1. Introduction

There are four fundamental forces in nature, namely:

(i) The strong nuclear force is the strongest among the four forces. The strong force is known to bind subatomic particles (such as protons and neutrons) to form nucleus.

(ii) The electromagnetic force which is in the order of $10^{-2}$ that of the strong force [1]. The electromagnetic force governs the interactions among electrically charged particles.

(iii) The weak nuclear force which is in the order of $10^{-14}$ of the strong force [1]. The weak force acts in each individual nucleons (i.e., collections of protons and neutrons) and is responsible for the radioactive decay when neutrons decay to protons and electrons.

(iv) The gravitational force which is the weakest among all forces. The gravitational force attracts any object with mass.

A field is a spatial distribution of quantity, which may or may not be a function of time [2]. To put it in simple terms, an electromagnetic field is basically the field produced as a consequence of positively and/or negatively charged particles, be at rest or in motion, and exerted forces among each other. The electromagnetic field consists of both the electric field and the magnetic field. During static condition, both electric and magnetic fields exist independently. When only an electric field is present and is constant in time, the field is known as an electrostatic field; similarly, when only a constant magnetic field is present, it is known as a magnetostatic field. When the fields change over time (i.e., in time-varying condition), however, both fields have to be concurrently present. This is to say that a time-varying electric field induces a time-varying magnetic field and vice versa [1], resulting in both fields being coupled together.

Due to its particle-wave duality nature, an electromagnetic field can be viewed as a continuous field which propagates in a wavelike manner, while at the same time, it can also be seen as quantized particles called photons. When the wave of the electromagnetic field propagates in an isotropic homogeneous medium, the electric and magnetic field components are mutually transverse to the direction of the energy transfer, as depicted in **Figure 1**. The radiation is therefore known as a transverse electromagnetic or TEM wave.

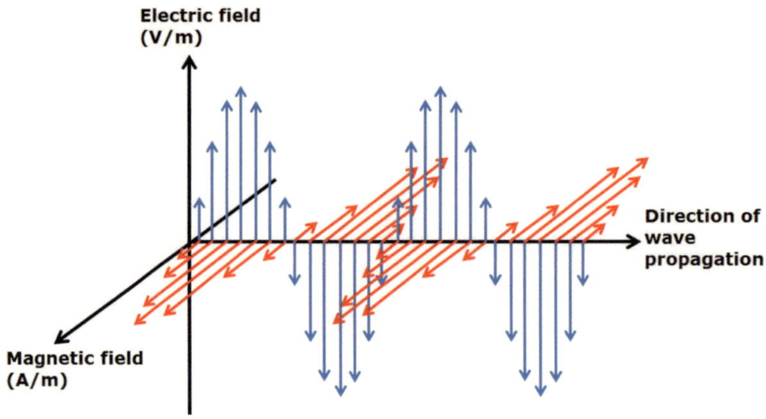


**Figure 1.**
*Transverse electromagnetic wave propagation.*

The distance between two adjacent troughs or crests of the electromagnetic wave is known as a wavelength. The wavelength is inversely proportional to the frequency of the wave (i.e., the tendency in which the wave repeats the same wave pattern). In other words, if the wave tends to repeat its cycle at a faster pace, the wavelength will become shorter. Likewise, if the pace of repetition decreases, the wavelength will become longer. The relationship between wavelength λ and frequency $f$ can be expressed as (1) below:

$$\lambda = \frac{1}{f\sqrt{\mu\epsilon}} \tag{1}$$

where $\frac{1}{\sqrt{\mu\varepsilon}}$ is the velocity of the wave propagation. Here, $\varepsilon$ and $\mu$ are, respectively, the permittivity and permeability of the medium where the wave propagates. The permittivity $\varepsilon$ measures the degree a material is polarized by the electric field, whereas the permeability $\mu$ dictates its ability in supporting the development of the magnetic field. Both electric and magnetic flux densities are, therefore, in direct proportion with $\varepsilon$ and $\mu$, respectively. Together with the conductivity parameter, σ, which describes the ease at which a charge can move freely in a material, these three parameters (i.e., $\varepsilon$, $\mu$, and σ) are referred to as the constitutive properties of the material.

## 2. Electromagnetic spectrum

Electromagnetic waves propagate at different frequencies. The electromagnetic spectrum is the classification of the waves in accordance to their range of frequencies. The classification is necessary since waves which radiate at different frequencies may be generated by different sources and may exhibit different effects on matters. The names given to the bands of frequencies may be broadly classified, in the order of decreasing wavelength and increasing energy and frequency, as follows:

(i) radio wave and microwave

(ii) terahertz wave

(iii) infrared radiation

(iv) visible light

(v) ultraviolet radiation

(vi) X-rays

(vii) gamma rays

A summary of the spectrum is given in **Figure 2**. It is to be noted that the frequencies at each edge of the band are merely approximations. The boundaries of each band are difficult to be defined since the waves are continuous and the bands may fade into each other.

### 2.1 Radio wave and microwave

The radio wave band ranges from around 3 Hz to 300 GHz. At the lower end of the radio wave (i.e., from 3 Hz to 300 MHz), the band is typically designated for radio and television broadcasts and is commonly known as the radio frequency RF signal. At the higher end (i.e., from 300 MHz to 300 GHz), the signal is used for telecommunication, satellite communication, food heating, wireless networking, wireless power transmission [3, 4], etc. Due to its vast applications at the higher end, this range of frequencies is specifically denoted as microwave. The EHF band which spans from 30 to 300 GHz holds important spectral and spatial information in the field of astrophysics [5]. The cosmic microwave background (CMB) radiation, for instance, peaks in the frequency range of 100–300 GHz [6]. The study of the CMB provides an in-depth understanding of the physics of the Big Bang and

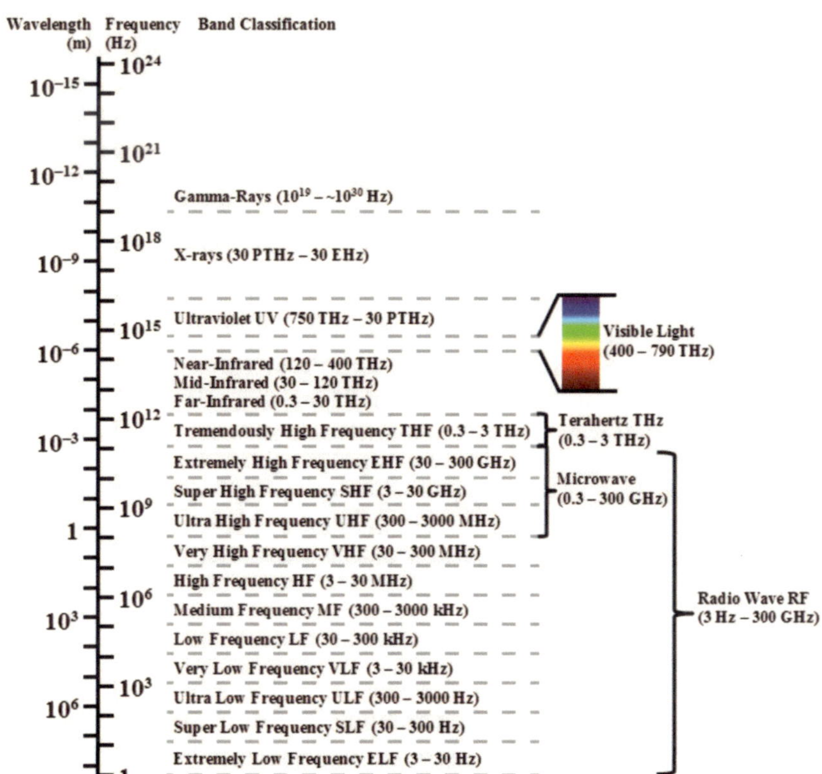


**Figure 2.**
*The electromagnetic spectrum.*

the formation of the early universe. Since this range of frequencies corresponds to wavelengths from 1 to 10 mm, it is also sometimes called the millimeter band.

Information is modulated in the radio wave signal and is emitted from a transmitter circuit to a receiver circuit, via antennas. In order to mediate effective communication, the antennas are required for [7]:

(i) efficient signal coupling and radiation

(ii) impedance matching, so as to minimize reflection when coupling the signal to the transmission line

Some of the common types of antennas used for RF communication are of the dipole [8–10], monopole [11–13], patch [14–17] and loop [18–21] configurations.

## 2.2 Terahertz wave

Sandwiched between the microwave and optical bands and falling within the $10^{12}$ Hz regime, the tremendously high-frequency (THF) band is more popularly known as the terahertz (THz) band. Since its wavelengths lie in the range of 0.1–1 mm, this band is therefore referred to as the submillimeter band as well.

Compared to microwaves, THz radiation has a shorter wavelength, and, thus, it possesses more energy to penetrate deeper and make sharper images. The radiation also scatters less than visible and near-infrared frequencies [22]. Unlike X-rays and ultraviolet (UV) light, THz waves are nonionizing. Hence, the radiation does not impose detrimental implications on living tissues [22]. Besides, the following characteristics are also shown when a material is illuminated with THz waves: polar liquids, such as water, absorb strongly in THz radiation, and metals are opaque to such radiation, whereas non-metals such as plastics, paper products, and non-polar substances are transparent [22]. Dielectrics, on the other hand, have characteristic absorption features peculiar to each material [22]. Due to the unique properties of this wave, THz waves are useful in an increasingly wide variety of applications, such as biology and medical sciences, homeland security, quality control of food and agricultural products, global environmental monitoring, etc. [23].

The THz region in the electromagnetic spectrum is of considerable importance in astrophysics. Like the millimeter band, the submillimeter band consists of spectral and spatial information of cosmic sources [5, 24]. The cold material (10–30 K) associated with the early stages of star and planet formation, as well as the earliest stages of galaxy formation, has its peak emission in the millimeter and submillimeter range [25]. By analyzing and mapping the lines in the THz band, it is possible to build complete models of astrophysical objects, which include temperature, density, large-scale movement of material, magnetic field strengths, isotope abundance, etc. [6].

Parabolic reflector antennas are used to detect THz waves in radio telescopes [24, 26, 27]. Due to the skin effect, transmission lines become very lossy at THz frequencies. Hence, waveguides are used instead to couple the signal and to channel it to the receivers [28–30].

## 2.3 Infrared radiation

The infrared IR radiation ranges from around 300 GHz to 400 THz and constitutes about 50% of the total sunlight. Since objects which radiate or sensors which detect IR radiation are only restricted to certain limited bandwidth, the band is further subdivided into three smaller parts, i.e., the far-infrared (FIR) (300 GHz–30 THz), mid-infrared (MIR) (30–120 THz) and near-infrared (NIR) (120–400 THz) sections.

The far-infrared (FIR) section behaves somewhat similar to the THz band. Waves in this band can be absorbed by the movements of molecules. Celestial sources such as cold clouds of gas and dust emit FIR waves. Detection of the FIR radiation can therefore be used to observe the formation of protostar in these clouds.

The mid-infrared (MIR) section is strongly absorbed by the earth atmosphere—particularly, by water vapor and carbon dioxide. It is also susceptible to molecular vibrations. This spectrum is also essentially important in astronomy. Planets absorb heat from the sun and reradiate it in the MIR spectrum. Also, interstellar dust and protoplanetary disks emit strongly in the MIR region. The 'heat-seeking' missiles used by the military are also designed to operate within this range. Black bodies, such as the heat from human bodies, radiate intensely in the MIR and FIR regions. Hence, the FIR and MIR signals are also used for thermal imaging.

The near-infrared (NIR) section is the closest to the lower end of the visible light frequencies. Hence, it is typically used in fiber optics communication. In astronomy, cooler stars such as the red giant stars and red dwarfs radiate very intensely at the NIR spectral region. Night vision devices, image sensors and NIR spectroscopy which finds wide applications in the medical field also make use of signals which oscillate at this range.

## 2.4 Visible light

The visible light is the most familiar spectrum to humans, and it constitutes about 40% of the total sunlight. Ranging from approximately 400 to 790 THz, this spectral region is the most sensitive to human eyes. Molecules and atoms absorb or release energy in this range of frequencies, allowing electrons to move to different energy levels. The change of bond structure in the retina of human eyes allows this spectrum to be visible to humans. Unlike signals from the radio wave to the infrared spectrum where radio telescopes are built to detect the signals, optical telescopes are used to visualize celestial objects which radiate at this spectrum. Visible light also excites chemical reactions in plants, mediating the process of photosynthesis.

## 2.5 Ultraviolet radiation

As the frequency increases above those of the visible light, the energy carried by the photons becomes very energetic—so much so that it can ionize atoms and disrupt molecular bonds. This is to say that electromagnetic waves in the ultraviolet (UV), X-rays and gamma rays bands cause ionizing radiations and that they could be detrimental to living things as a whole and human beings in particular.

UV radiation ranges approximately from 750 THz to 30 PHz, and it constitutes almost 10% of the total sunlight. The spectrum can be further divided into the ultraviolet A (UVA) section which spans roughly from 750 to 950 THz, ultraviolet B (UVB) section which spans roughly from 950 to 1000 THz, and ultraviolet C (UVC) section which spans roughly from 1 to 30 PHz. The UVC emission carries the highest energy and is the most hazardous. But it is literally absorbed by the ozone and atmosphere layers before reaching the earth. Likewise, the UVB emission is also absorbed mostly by the stratospheric ozone layer. Hence, only the UVA emission and a very small fraction of the UVB emission reach the earth surface. The UVA emission carries the lowest energy and is the least damaging of all. Besides the sun, UV radiation can also be artificially produced, such as from electric arc, gas discharge lamps, etc.

Due to its ionizing behavior, UV light is known to be the main cause of skin cancer and skin burn. The high energy level carried by the wave at this spectral range may damage the DNA molecules, resulting in the formation of thymine dimers. In order to prevent from being directly exposed to the UV radiation, sunscreens are usually used. The active ingredients in sunscreens can either be organic (such as dioxybenzone and oxybenzone) or inorganic compound (such as titanium dioxide and zinc oxide).

Although being constantly exposed to UV radiation is harmful, moderate exposure may actually be beneficial to the human body. This is because UV radiation mediates the production of vitamin D in the human body. Vast applications may also be found using UV radiation. In the biomedical field, UV light is used in phototherapy to treat severe skin problems such as psoriasis and eczema. The photolithography process in the semiconductor industries [31] also uses UV radiation to print circuits onto the photoresists deposited on wafers. Since fluorescent dyes and certain fluids illuminate brightly under UV light, the radiation is also used in the security and forensic areas. In astronomy, wave signals at this range are detected to analyze the composition of interstellar and intergalactic medium. However, since UV is greatly absorbed by the ozone layer, the UV telescope has to be placed in space. An example of such telescope is the Hubble Space Telescope (HST) [32].

## 2.6 X-rays

X-rays which were also named as the X-radiation by its discoverer, Wilhelm Röntgen, have frequencies much higher than the UV radiation, which ranges around 30 PHz–30 EHz. X-rays with higher energy levels are known as hard X-rays, whereas those with lower energy levels are known as soft X-rays. Since soft X-rays are susceptibly absorbed in air, it is less useful than hard X-rays. Hence, the term 'X-rays' is generally referred to hard X-rays when they are being applied.

X-rays can be generated via synchrotron radiation, X-ray fluorescence or bremsstrahlung, and they can be detected via imaging detectors which uses photographic films.

X-rays have significantly shorter wavelengths than the visible light. They can therefore penetrate most tissues in the human body. When they propagate through the human body, different amount of the energy is absorbed by different types of tissues. The absorption rate depends on the radiological density of the tissues. Bones which has high radiological density absorbs the energy considerably more than tissues with lower densities, such as muscle, fat and air-filled cavity in the lungs. When images are produced using X-rays, the parts with bones will appear whitish, whereas those with softer tissues will appear shadowy. Because of this reason, X-rays are widely used for medical imaging, such as radiography and computed tomography (CT) scanning. The employment of X-rays in luggage scanning in airport security is based on a similar concept as medical imaging. Besides imaging, X-rays are also used in radiation therapy for cancer treatments in the medical field. Since some of the cosmic sources emit waves in this spectral range, it is therefore useful in the field of astrophysics—the detection of which, like the case of UV radiation, is conducted using orbiting telescopes. Also, in a technique called X-ray crystallography, X-rays are applied to study the structures of crystals.

Like UV radiations, X-rays are highly ionizing and may cause deleterious effects on humans. Hence, the applications of X-rays are often conducted in lead-shielded rooms, and the practitioners are to wear lead aprons and gloves. The amount of X-ray exposure is measured using an ionization chamber, and the dosage of X-rays, in which the user is exposed to, is measured using a dosimeter.

## 2.7 Gamma rays

The distinction between X-rays and gamma rays is not unanimous. One general definition to distinguish between these two is that X-rays are generated via the acceleration of electrons, whereas gamma rays are generated from the radioactive decay of atomic nuclei. However, this definition may not always be valid since the source of radiation could not always be identified. The lowest frequency of the gamma rays is about $10^{19}$ Hz, while the highest known frequency that has been detected hitherto reaches about $10^{30}$ Hz.

Gamma rays are generated from the following processes:

(i) Nuclear fusion. Nuclear fusion is a natural phenomenon in which hydrogen nuclei fuse into helium nuclei, releasing energy in the form of gamma rays. Stars, such as the sun in the Milky Way galaxy, are powered by nuclear fusion. The creation of new elements in this process is known as stellar nucleosynthesis.

(ii) Nuclear fission, which is pretty much the reverse process of nuclear fusion. In nuclear fission, a nucleus splits into smaller and lighter nuclei. In the process of doing so, the energy is released as gamma rays.

(iii) Alpha decay. In alpha decay, a nucleus emits an alpha particle, and it, subsequently, reduces its atomic number by 2 and its mass by 4. The excess energy is emitted as gamma rays.

(iv) Gamma decay. When a nucleus contains too much energy, the energy is emitted as gamma ray photons. Since there is no particle released in the process, the charge and mass compositions of the nucleus remain unchanged.

Shielding materials such as lead are to be used for protection since gamma rays can cause the mutation of genes, resulting in the contraction of cancer. Quite ironically, however, gamma rays are also used to kill malignant cancerous tumors. In gamma knife radiosurgery, surgeons focus gamma ray beams toward the targeted region to kill the cancerous cells. Likewise, gamma rays are also used in food irradiation to kill microorganisms and to sterilize medical equipment. In astrophysics, gamma rays are detected to study pulsars, quasars, nebulae, and gamma ray bursts (GRBs).

## 3. Maxwell's equations

The discoveries made by scientists over the past millennia have contributed towards the profound understanding of electromagnetics that we have today. Among these scientific discoveries, it is the experimental observations reported independently by Ampere, Faraday and Gauss which inspired James Clerk Maxwell to establish the unified theory of electricity and magnetism. In 1873, Maxwell published his formulations in his textbook *A Treatise on Electricity and Magnetism*. The complexity of the formulations in the textbook was later reduced by Oliver Heaviside in 1881 to four sets of differential equations.

More popularly referred to as Maxwell's equations today, these four sets of notable mathematical equations which outline the fundamental principles of electromagnetism are tabulated in **Table 1**. Eq. (2.1) in the table describes the observation reported by the English physicist, Michael Faraday. According to Faraday, when the magnetic field intensity (**H**) or magnetic flux density ($\mu\mathbf{H}$) varies with time ($t$), a force $V = \oint_C \mathrm{E} \cdot dl$ will be induced, where **E** is the electric field intensity and the line integral is performed over a line contour $C$ bounding an arbitrary surface. Established by the French physicist Andre-Marie Ampere, Ampere's circuital law in Eq. (2.2) states that the circulation of **H** around a closed loop will result

| Differential form | Integral form | Law | |
|---|---|---|---|
| $\nabla \times E = -\mu \frac{\partial H}{\partial t}$ | $\oint_C E \bullet dl = -\int_S \mu \frac{\partial H}{\partial t} \bullet dS$ | Faraday's law | (2.1) |
| $\nabla \times H = J + \varepsilon \frac{\partial E}{\partial t}$ | $\oint_C H \bullet dl = \int_S \left(J + \varepsilon \frac{\partial E}{\partial t}\right) \bullet dS$ | Ampere's law | (2.2) |
| $\nabla \bullet E = \frac{\rho}{\varepsilon}$ | $\oint_S E \bullet dS = \int_V \frac{\rho}{\varepsilon} \bullet dV$ | Gauss' law for electricity | (2.3) |
| $\nabla \bullet H = 0$ | $\oint_S H \bullet dS = 0$ | Gauss' law for magnetism | (2.4) |

**Table 1.**
*Maxwell's equations.*

in current traversing through the surface bounded by the loop. Here, **J** is the convection current density, while $\varepsilon \frac{\partial E}{\partial t}$ is the displacement current density. Eqs. (2.3) and (2.4) are formulated based on the observations made by the German physicist, Carl Friedrich Gauss. According to Gauss, the total electric flux density (ε**E**) flowing out from an enclosed surface *S* is equivalent to the total charge (ρ) encapsulated within *S*, i.e. (2.3), whereas the net outward **H** from *S* is invariably zero, i.e. Eq. (2.4).

## Author details

Kim Ho Yeap[1*] and Kazuhiro Hirasawa[2]

1 Universiti Tunku Abdul Rahman, Jalan Universiti, Bandar Barat, Kampar, Perak, Malaysia

2 Universiti of Tsukuba, Tsukuba, Ibaraki, Japan

*Address all correspondence to: yeapkimho@gmail.com

## References

[1] Ulaby FT. Electromagnetics for Engineers. Upper Saddle River, New Jersey: Pearson; 2005

[2] Cheng DK. Field and Wave Elecromagnetics. 2nd ed. US: Addison-Wesley; 1989

[3] Nakayama M, Hiraguri T, Mitsui T, Nishimori K, Matsuda T, Hirasawa K, and Ho YK. Proposal of multidirectional power transmission system using monopole antennas and parasitic elements. In: Proceedings 2017 IEEE AP-S Symposium on Antennas and Propagation and USNC-URSI Radio Science Meeting. San Diego, CA, USA: IEEE; 2017

[4] Cheah ARC, Yeap KH, Hirasawa K, Yeong KC, and Nisar H. Optimization of a wireless power transmission system. In: Proceedings of the International MultiConference of Engineers and Computer Scientists; Tsim Sha Tsui, Hong Kong: IAENG; 2016. pp. 590-592

[5] Yeap KH, Tham CY, Yeong KC, Chong KH, Rizman ZI, Yang CC. Analysis of normal and superconducting coplanar waveguides in radio astronomy. International Journal of Electronics, Computer, and Communications Technologies. 2011;**2**:9-12

[6] Withington S. Terahertz Astronomical Telescopes and Instrumentation. Philosophical Transactions. UK: Royal Society of London; 2003. pp. 395-402

[7] Sadiku MNO. Elements of Electromagnetics. 4th ed. New York: Oxford; 2007

[8] Hirasawa K, Haneishi M. Analysis, Design and Measurement of Small and Low-Profile Antennas. US: Artech House; 1991

[9] Morioka T, Hirasawa K. MoM calculation of the properly defined dipole antenna factor with measured Balun characteristics. IEEE Transactions on Electromagnetic Compatability. 2001;**53**:233-236

[10] Yeap KH, Cheah ARC, Hirasawa K, Yeong KC, Lai KC, Nisar H. Optimization of wireless power transmission systems with parasitic wires. ACES Journal. 2017;**32**:806-812

[11] Chen ZN, Hirasawa K, Wu K. A novel top-sleeve monopole in the two parallel plates. IEEE Transactions on Antennas & Propagation. 2012;**49**:438-443

[12] Chen ZN, Hirasawa K, Wu K. Numerical analysis of sleeve monopole in parallel-plate waveguide. International Journal of RF Microwave & Computer-Aided Engineering. 2012;**11**:86-98

[13] Chen ZN, Hirasawa K, Wu K. A broad-band sleeve monopole integrated into parallel-plate waveguide. IEEE Transactions on Microwave Theory & Technique. 2000;**48**:1160-1163

[14] Cheong HR, Yeap KH, Lai KC, Teh PC, Nisar H. A compact CPW-fed antenna with S-shaped patches for multiband applications. Microwave and Optical Technology Letters. 2017;**59**:541-546

[15] Yeo WL, Lai KC, Yeap KC, Teh PC, Nisar H. A compact-dual band hook-shaped antenna for wireless applications. Microwave and Optical Technology Letters. 2017;**59**:1882-1887

[16] Voon CS, Yeap KH, Lai KC, Seah CK, Nisar H. A compact double-psi-shaped dual band patch antenna for WLAN/LTE applications. Microwave and Optical Technology Letters. 2018;**60**:1271-1275

[17] Yeap KH, Ismail W, Yeap KH. Analytical model for E-shaped

microstrip patch antenna. ACES Journal. 2017;**32**:332-338

[18] Hiraguri T, Ojiro Y, Hirasawa K, Shouno K. Polarization control of a loop antenna by PIN diodes. IET Microwaves, Antennas & Propagation. 2007;**1**:592-596

[19] Sumi M, Hirasawa K, Shi S. Two rectangular loops fed in series for broadband circular polarization and impedance matching. IEEE Transactions on Antennas and Propagation. 2004;**51**:551-554

[20] Morishita H, Hirasawa K, Nagao T. Circularly polarised rhombic hula hoop antenna. Electronics Letters. 1996;**32**:946-947

[21] Morishita H, Hirasawa K, Nagao T. A circularly polarized broadband rhombic loop antenna. IEICE Transactions on Communications. 1996;**E79-B**:865-870

[22] Pospiech M, Kaiserslautern TU. Terahertz Technology: An Overview. Department of Physics and Astronomy. UK: University of Sheffield; 2003

[23] Tonouchi M. Cutting-edge terahertz technology. Review article. Nature Photonics. 2007;**1**:97-105

[24] Yeap KH, Tham CY. Optimization of an offset receiver optics for radio telescopes. Journal of Infrared, Millimeter, and Terahertz Waves. 2018;**39**:64-76

[25] Barychev AM. Superconductor-insulator-superconductor THz mixer integrated with a superconducting flux-flow oscillator [Ph.D. thesis]. Netherlands: Delft University of Technology; 2005

[26] Yeap KH, Yiam CY, Lai KC, Loh MC, Lim SK, Rizman ZI. Analysis of offset antennas in radio telescopes. International Journal on Advanced Science, Engineering, Information and Technology. 2016;**6**:997-1004

[27] Yeap KH, Loh MC, Tham CY, Yiam CY, Yeong KC, Lai KC. Analysis of reflector antennas in radio telescopes. Advanced Electromagnetics. 2016;**5**:32-38

[28] Yeap KH, Wong EVS, Nisar H, Lai KC, Ng CA. Attenuation in circular and rectangular waveguides. Electromagnetics. 2017;**37**:171-184

[29] Yeap KH, Ong SS, Nisar H, Lai KC, Ng CA. Attenuation in superconducting circular waveguides. Advanced Electromagnetics. 2016;**5**:34-38

[30] Yeap KH, Teh JSM, Nisar H, Yeong KC, Hirasawa K. Attenuation in superconducting rectangular waveguides. Frequenz Journal of RF-Engineering and Telecommunications. 2015;**69**:111-117

[31] Ahmad I, Ho YK, Majlis BY. Fabrication and characterization of a 0.14 μm CMOS device using ATHENA and ATLAS simulators. International Scientific Journal of Semiconductor Physics, Quantum Electronics, and Optoelectronics. 2006;**9**:40-44

[32] Tham CY. Space is Curved, the Universe is Flat. Malaysia: Malaysiakini; 2002

## Section 2

# Electromagnetic Waves Sources

Chapter 2

# Vacuum Microwave Sources of Electromagnetic Radiation

*Gennadiy Churyumov, Jinghui Qiu and Nannan Wang*

## Abstract

This chapter contains new simulation results concerning the physical foundations of how microwave tubes operate based on Cherenkov's mechanism of radiation (interaction with slow electromagnetic wave) and some experimental results connected with improving the output characteristics of the magnetrons (a mm band surface wave magnetron and a magnetron with two RF outputs of energy), as well as results of computer modeling of a 320-GHz band traveling wave tube (TWT). The results of analytical calculations and computer modeling, a phase bunching process in the mm band surface wave magnetron, are considered. It is shown that the process of phase focusing has two features associated with a concentration of RF wave energy close to the vanes of an anode block and higher electron hub of a space charge as compared to the classical magnetrons. The features and examples of practical application of the magnetron with two RF outputs of energy are presented. It is shown that the main advantage of the magnetrons is its extended functionalities (for example, possibility of frequency tuning including electronic tuning of a frequency from a pulse to pulse). The presented materials will be of interest not only for starting researchers but also for those who have microwave tube experience.

**Keywords:** Cherenkov's radiation, electromagnetic field, electron beam, magnetron, TWT, frequency tuning, millimeter range, terahertz range

## 1. Introduction

The applications of electromagnetic fields are of great importance in such areas as radar and navigation, communications, information and communication technologies, industry and agriculture, medicine, etc.

It is the interaction of moving charged particles (for example, electrons) with the electromagnetic field that is the cornerstone of electromagnetic phenomena. It is known that the electron moving rectilinearly and evenly with constant speed does not radiate. Moreover, the electromagnetic field, which exists around particle, moves together with particle at the same speed, and its properties remain invariable. If the nature of movement of a particle and/or its speed changes, for example, the trajectory of movement becomes curvilinear or the electron begins to move unevenly (turns to be accelerated or slowed down), the state of its own electromagnetic field also changes. As a result, there arises a free electromagnetic field, i.e., electromagnetic radiation (EMR), which has wave nature and freely advances in the environment in the form of electromagnetic wave. Depending on the existing conditions of its propagation (or accumulation of electromagnetic energy), which are determined by the properties of the medium (for example, features of the spatial configuration of periodic waveguide and resonant electrodynamic structures) as well as the character of movement trajectories of electrons in an electron beam, there might exist the EMR different types.

The following types of radiation are known: Cherenkov's radiation (interaction with slowed-down waves), transient and diffraction radiations (or Smith-Pascrell's radiation as applied to optical band), and bremsstrahlung and magnetodeceleration radiation (while having a magnetic field), as well as its versions: synchrotron or undulator radiation (for a relativistic case) and cyclotron radiation (for a case of the movement of nonrelativistic particles) [1–7]. The analysis reveals that various mechanisms of the radiations have both common and specific features of their occurrence. It is also necessary to note that the existence of one dominating type of radiation may result in the interference of several types of radiation under certain conditions [8]. As a result, the output parameters of a microwave source will depend on the fact how efficiently the conditions of electron interaction will be correlated with an electromagnetic field from the viewpoint of conversion efficiency of the energy, being reserved in an electronic flow, into electromagnetic energy, and how fully and correctly the requirements to the devices distributing and accumulating an electromagnetic field (periodic and resonator electrodynamic structures) are formulated.

Studying the features of physics of the electromagnetic radiations has led to producing various microwave vacuum tubes in the wide range of frequencies [9]. Considerable recent attention has been focused on creating microwave sources in a short-wave part of millimeter (0.1–0.3 THz) and terahertz (0.3–3.0 THz) ranges. It is known that the absolute advantage of terahertz range is the broad band of frequencies, as well as the ability to penetrate through opaque media, including metals, organic materials, etc. This property positively distinguishes it from ionizing radiation (for example, X-rays) while opening wide perspectives to diagnose a variety of diseases in medicine [10].

The scheme of practical mastering of a short-wave part of a spectrum of electromagnetic oscillations is shown in **Figure 1**.

It is evident nowadays that the lack of the microwave vacuum devices with continuous power output from a few watts to tens and more watts in this part of the electromagnetic spectrum restricts the opportunities of further development and improvement of technologies in such areas as spectroscopy, radio astronomy, space, and biochemical research, as well as producing a new generation of information and communication systems. Besides the development of this frequency range will allow intensifying safety control (search and detection of explosives, remote identification of chemical substances) to exercise production quality control of finished goods (checking packages of medical products, etc.). The application of relativistic tubes (for example, the relativistic magnetrons or the fast-wave devices like free-electron lasers or gyrotrons) for solving the above-mentioned

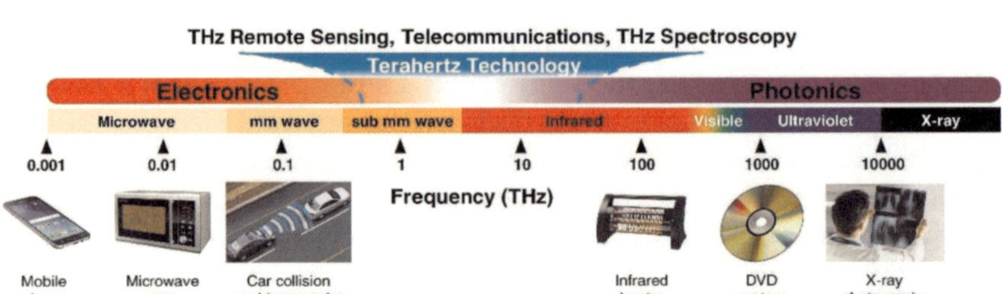


**Figure 1.**
*The scale of electromagnetic waves and areas of their application.*

problems allows providing the needed levels of output power in this range. However, these devices possess large mass-dimensional characteristics and require for operating the high values of voltages and strong magnetic fields. In this regard, producing effective and compact microwave sources of electromagnetic oscillations in this frequency range seems to be much more attractive with the help of miniaturizing the constructions of the classical microwave tubes possessing Cherenkov's radiation mechanism.

The present chapter deals with vistas of developing the microwave sources of electromagnetic oscillations (the magnetron and the O-type TWT) whose operation is based on the interaction of an electron beam with the slowed-down electromagnetic wave of electrodynamic structure ($\nu_{ph} < c$, where $\nu_{ph}$ is the phase velocity of electromagnetic wave and $c$ is the light speed), i.e., there exists Cherenkov's radiation mechanism.

## 2. Microwave tubes with Cherenkov's radiation mechanism

### 2.1 General subjects

Microwave sources operating on basis of Cherenkov's radiation are a wide class of microwave tubes, including the magnetrons, backward-wave tubes (BWTs), resonant TWTs, and orotron [9]. As for now, microwave sources with Cherenkov's interaction have allowed to reach the maximum levels of peak capacity ~3 GW in the 3-cm range of wavelength and more than 5 GW in the 8-cm range at a pulse duration of 1–10 nanoseconds including the generation of ultrashort pulses of electromagnetic radiation (the effect of superradiation) [11]. Let us consider the development of these tubes on the example of their most famous representatives, which are magnetrons and TWTs.

### 2.2 Magnetron

Historically, the first microwave tube whose operation is based on Cherenkov's interaction was a magnetron [12]. The successful combination of properties of a multimode oscillating system of the magnetron and electronic processes occurring in its interaction space has allowed the magnetron to become one of the most effective microwave generators [13, 14]. The constructions of magnetrons developed nowadays generate electromagnetic oscillations in the frequency range from 300 MHz to 300 GHz. The output power of continuous magnetrons ranges from a few fractions of a watt to several tens of kilowatts, and the pulsed magnetrons generate the electromagnetic oscillations with output power from 10 W to 20 MW. Both pulsed and continuous magnetrons are widely used in different ground-based and on-board electronic systems, industrial and microwave household heating systems, physical experiments for plasma heating, and the acceleration of charged particles, as well as in the phased antenna grids and space systems of solar energy conversion to direct current energy (the system of wireless energy transmission to the earth). The miniaturization of the magnetron construction has allowed to reduce their weight to 100 g at the pulsed power of 1 kW and the efficiency of which achieves about 50% that is quite competitive with the best samples of modern transistor oscillators [10, 13–20].

Concerning the existing trends of investigations, it is necessary to emphasize the problems of improvement of the cathode (emitter) operation of magnetrons. It is found (see, e.g., [13, 14, 21]) that the changes and instability of the cathode emission characteristics influence considerably the frequency stability of the magnetron. On the other hand, the disorder in the operation mode of the cathode connected with its overheat leads to essential reduction of durability of the cathode and the cathode node as a whole. In this regard, it is of great interest to study the problem of excess electronic and/or ionic bombing of the cathode and to clarify the role of turbulence of a cathode electronic cloud in the course of secondary-emission multiplication of an electronic beam [22, 23]. In order to improve the magnetron operation, there conducted active investigations of new effective cathode materials and cathode node constructions, providing high durability and emission stability. It especially concerns investigations connected with applying the cold secondary-emission cathodes providing virtually instant readiness and nonfilament ("cold") start of magnetrons [22, 24–28].

The scope of magnetrons continues to expand constantly that is primarily caused by their advantages such as high electronic efficiency (more than 80%), relatively low voltages (in particular, anode voltage), the high relation of power output level to the weight of the tube, the compactness of construction, the simplicity of production, and comparatively low cost [10, 15]. However, such drawbacks of magnetrons as the low stability of frequency, the increased noise levels, and spurious oscillations require carrying out additional investigations and analyzing the ways for improving its output characteristics, in particular, in a short-wave part of a millimeter range. Solving mentioned above problems will allow to increase competitiveness of magnetrons and to expand their functionality in comparison with other microwave tubes, such as one-beam and multibeam klystrons, klystrons, complexificated microwave sources on the basis of "the solid-state generator and TWT" chains, etc. [10, 13, 14, 21, 22, 24, 25].

### *2.2.1 The surface wave magnetrons*

The state-of-the-art evolution of magnetrons is associated with increasing the frequency (phase) stability and rising the lifetime, as well as enhancing reliability due to an application of the cold cathodes [21, 24, 26–28]. In particular, there is a significant interest in the development of magnetrons in the millimeter range. These magnetrons can be applied in radar systems [29]. Among possible designs of similar magnetrons, it is necessary to select the magnetrons working on higher space harmonics (for example, by using as operating a first negative ($-1$) space harmonic or an oscillation does not $-\pi$ mode [30]). The magnetrons operating in such modes are named the surface wave magnetrons [30, 33]. According to the approach described in [32], a method has been developed to calculate the parameters and the operation modes. It is necessary to note that a major feature concerning to the application of the surface wave magnetron is associated with the generation of electromagnetic waves in the millimeter range at a considerably low magnetic fields and increased sizes of an interaction space. The prototypes of the surface wave magnetrons have been built. The prototypes operate at the $\frac{\pi}{2}$ – mode, and they provide the following output parameters: wavelength band from 1.25 mm up to 6.8 mm, a level output pulsed power from 1 up to 150 kW, and efficiencies 0.8–20% [31].

In spite of the intensive research over the years and getting experimental results including the constructions of the surface wave magnetrons, up to now, we have no necessary and full information about physical processes occurring in the interaction space of given magnetrons. Therefore, for studying and understanding the features of a mechanism of nonlinear interaction into an interaction space of the magnetron, it is necessary to carry out additional computer modeling, a phase bunching process of an electron beam under its interaction with surface wave by using the Particle-in-Cell Method.

For studying, we used the design of a 3-mm range magnetron. Schematically, this design presents in **Figure 2**. The essential geometry sizes of an interaction space of this magnetron are presented in **Table 1**. As the operation mode, the mode other than the $\pi-$ mode was taken, namely, the oscillation $\frac{\pi}{2}-$ mode. As cathode of the magnetron, an indirectly heated oxide cathode, which produced an emission current density of $\sim$2.0 A/sm$^2$, was used [37].

For computer modeling, we used the 2-D mathematical model of the magnetron described in [34]. The basis of this model is the self-consistent set of equations including the motion equation, the equation of excitation, and Poisson's equation for calculation of space charge forces.

As already mentioned above, the theoretical basis of the surface wave magnetrons was described in the works [30–32]. It is necessary to note that the distinctive feature of electron-wave interaction in given magnetrons (for example, as distinguished from the classical magnetrons [14], using the $\pi-$ mode as the operation one) is the distribution of an electromagnetic wave in the neighborhood of a surface of the RF structure and its interaction with electrons on a top of the space charge hub. In order to understand the mechanism of interaction of an electromagnetic field with an re-entrant electron beam, it is very important to define the features of the radial and azimuthal distributions of the electromagnetic wave in an interaction space (between a cathode and inside surface of an anode block (see **Figure 2**), as well as to provide clearer understanding about behavior of electrons and their motion trajectories.

An interaction space of the magnetron is shown in **Figure 2**, schematically. In order to determine the electromagnetic field in the resonance RF structure (anode

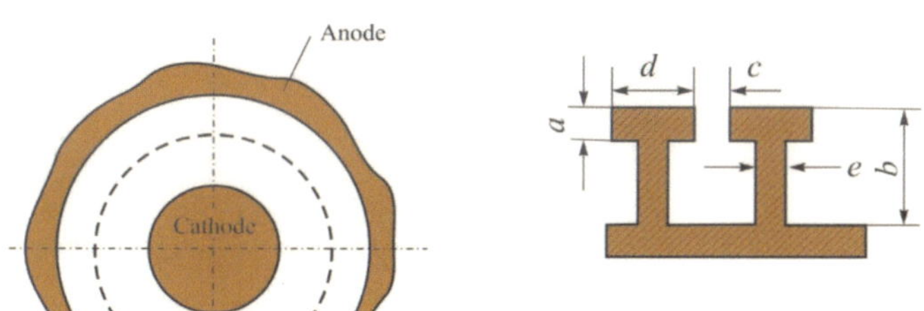


| a, mm | b, mm | c, mm | d, mm |
|---|---|---|---|
| 0,1 | 0,35 | 0,1 | 0,335 |

**Figure 2.**
*Schematically image of a surface wave magnetron.*

| Parameters | Symbol | Unit | Value |
|---|---|---|---|
| 1. Frequency | $f$ | GHz | 9.2 |
| 2. Cathode radius | $r_c$ | mm | 0.85 |
| 3. Anode block radius | $r_a$ | mm | 1.665 |
| 4. Anode block height | $h$ | mm | 4.0 |
| 5. Resonators number | $N$ | mm | 24 |

**Table 1.**
*The main parameters of surface wave magnetron.*

block), we used the decision obtained from Maxwell's equations for free space without charged particles [14]. As a result, the expressions for components electromagnetic field in the interaction space may be written as

$$E_\phi(r,\phi) = E_m \frac{N\theta}{\pi} \sum_{m=-\infty}^{\infty} \left(\frac{\sin\gamma\theta}{\gamma\theta}\right) \times \frac{Z_\gamma{}'(kr)}{Z_\gamma{}'(kr_a)} \cdot e^{j\gamma\phi}; \tag{1}$$

$$E_r(r,\phi) = -jE_m \frac{N\theta}{\pi kr} \sum_{m=-\infty}^{\infty} \gamma \cdot \frac{\sin\gamma\theta}{\gamma\theta} \times \frac{Z_\gamma(kr)}{Z_\gamma{}'(kr_a)} \cdot e^{j\gamma\phi}, \tag{2}$$

where $k = 2\pi/\lambda$ is the propagation factor in free space; $\lambda = c/f$ the wavelength of the magnetron; $2\theta$—the angle subtended by a space between segments of the anode block;

$$Z_\gamma(kr) = J_\gamma(kr) - \frac{J_\gamma{}'(kr)}{N_\gamma{}'(kr_a)} \cdot N_\gamma(kr) \tag{3}$$

and

$$Z_\gamma{}'(kr) = J_\gamma{}'(kr) - \frac{J_\gamma{}'(kr_c)}{N_\gamma{}'(kr_a)} \cdot N_\gamma{}'(kr) \tag{4}$$

are the combinations of the well-known Bessel $J_\gamma(kr)$ and Neumann $N_\gamma(kr)$ functions; $\gamma$ is zero or any positive or negative integer.

In the expressions of Eqs. (1) and (2), we have the ratio of $\frac{Z_\gamma{}'(kr)}{Z_\gamma{}'(kr_a)}$ and $\frac{\gamma Z_\gamma{}'(kr)}{krZ_\gamma{}'(kr_a)}$, which are the structure functions of the electromagnetic field. From these expressions, we have seen that their values depend on a radius-vector $\vec{r}$ and bring about changing the magnitude of the $\gamma$-th mode in the radial direction of the interaction space. It is necessary to note that when $kr_a \ll \gamma$ (so-called long-wave approximation), the expressions can be simplified as [14]:

$$\Psi_r^\gamma(r) = \frac{Z_\gamma{}'(kr)}{Z_\gamma{}'(kr_a)} \approx \left(\frac{r}{r_a}\right)^{\gamma-1} \cdot \left[\frac{1-\left(\frac{r_c}{r}\right)^{2\gamma}}{1-\left(\frac{r_c}{r_a}\right)^{2\gamma}}\right]; \tag{5}$$

and

$$\Psi_\phi^\gamma(r) = \frac{\gamma Z_\gamma{}'(kr)}{krZ_\gamma{}'(kr_a)} \approx \left(\frac{r}{r_c}\right)^{\gamma-1} \cdot \left[\frac{1+\left(\frac{r_c}{r}\right)^{2\gamma}}{1-\left(\frac{r_c}{r_a}\right)^{2\gamma}}\right]. \tag{6}$$

Thus, the expressions obtained for components of an electromagnetic field of (1) and (2) in terms of (5) and (6) allow getting an electromagnetic field for the $\gamma$ – mode of a cold resonance anode block of a magnetron as

$$\vec{E}\left(\vec{r},t\right) = \mathrm{Re}\left\{\vec{E}\left(\vec{r}\right) \cdot e^{-j\omega_\gamma t}\right\}, \tag{7}$$

where $\vec{E}\left(\vec{r}\right) = E_r(r,\phi) \cdot \vec{r^0} + E_\phi(r,\phi) \cdot \vec{\phi^0}$, $\omega_\gamma = \omega_\gamma{}' - j\omega_\gamma{}''$—the complex cold frequency of the $\gamma$ – mode; $\omega_\gamma{}' = 2\pi \cdot f$—the angular frequency of the $\gamma$ – mode; $\omega_\gamma{}''$—the coefficient of attenuation.

Each mode excited in the anode block of the magnetron is characterized by certain distribution of the electromagnetic field and its frequency $\omega_\gamma = \omega_\gamma{}'$ in an approximation $\omega_\gamma{}'' = 0$. In the general case, the electromagnetic field in the interaction space is not sinusoidal and may be presented as a sum of the space harmonics, each of which corresponds to the wave rotating with an angular velocity $\Omega_\gamma$ and containing along the azimuthal length of the interaction space of a magnetron the whole number of the complete periods

$$\gamma = n + mN, \tag{8}$$

where $n = 0, 1, 2, \ldots, N/2$ is the number of the fundamental mode ($m = 0$); $m = \pm 1, \pm 2, \pm 3, \ldots$ is the integers corresponding to the high-order space harmonics. It is known (see, example, [13, 14]) that the excitation condition of the resonant system of the magnetron (or so-called a condition synchronism) may be written as

$$\Omega_e = \Omega_\gamma, \tag{9}$$

where $\Omega_e$ is the angular velocity of rotating electron spokes (or an electron beam closed on itself—re-entrant electron beam). On the other hand, we have an electron cloud, which circles in the interaction space around the cathode [9, 14], i.e., there is an additional condition, which is associated with a re-entrant electron beam and may be written as

$$\gamma = \frac{\omega_\gamma}{\Omega_e}. \tag{10}$$

The fundamental results of theoretical analysis are shown in **Figure 3**. The comparison of the radial functional dependences of the structural functions, Eq. (5) and Eq. (6), for two cases at using the higher space harmonics (for example, space harmonic—1 and $\gamma = 18$, curve 1) and at a classical $\pi$−mode ($m = 0$ and $\gamma = 12$, curve 2), shows that in the first case for effective interaction between electrons and electromagnetic wave, it is necessary to form the electronic hub having a height more than 0.85.

The trajectories of moving electrons as a result of interacting with electromagnetic field of the (−1) space harmonic are shown in **Figure 4**. Besides, here for comparison, we can see the trajectories of the electrons in the static mode of magnetron operation (dashed curves). It is seen that the phase focusing of the electron beam takes place in the range of the proper phases of the RF wave. In the range of the improper phases, we observe the multiplication secondary electrons process, increasing the density of space charge in the electron hub.

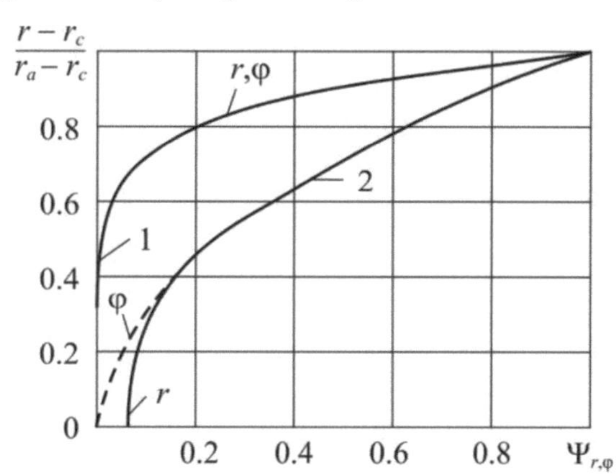


**Figure 3.**
*The radial distributions of the structural functions.*

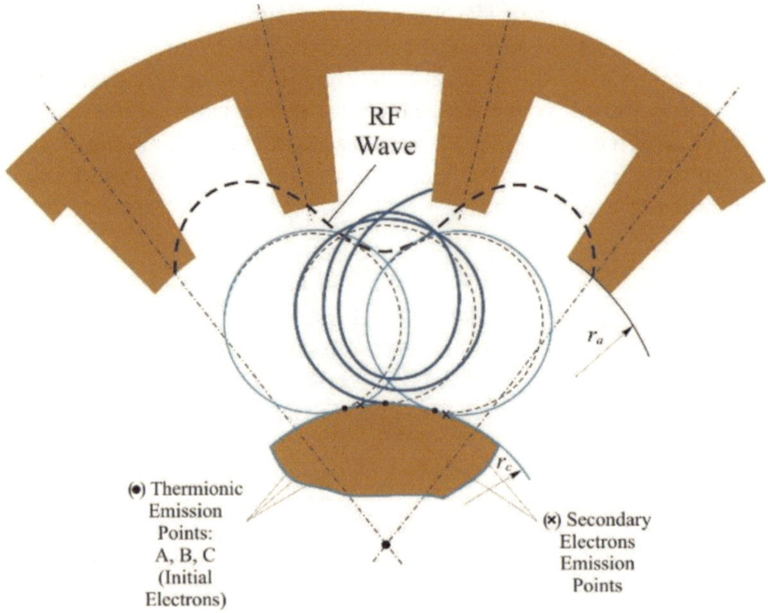


**Figure 4.**
*The trajectories of electrons in the interaction space [35].*

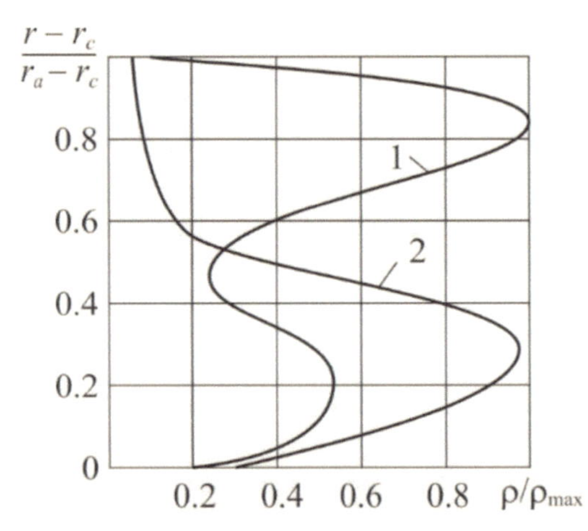


**Figure 5.**
*The radial distributions of space charge density in the interaction spaces of the surface wave magnetron (1) and the classical ($\pi$−mode) magnetron (2).*

**Figure** 5 shows the steady-state radial distributions of space charge densities in the interaction spaces of a surface wave magnetron and a classical magnetron. As can be seen, there is a fundamental difference between the two processes of the phase focusing. It is associated with available maximum of the space charge density in the immediate neighborhood of the surface of the RF structure (anode block). The availability of a second maximum of the space charge density allows the double stream state to be established in the electron hub.

It is also important to note that the operation mode on higher spatial harmonics applied in the magnetron, namely on −1 spatial harmonic or $\frac{\pi}{2}$−mode of oscillation, was rather successfully used for creating a relativistic prototype of the high-power magnetron with pulsed power of ~1 MW at the frequency of 37.5 GHz [34].

DOI: http://dx.doi.org/10.5772/intechopen.83734

On the basis of the design of the 3-mm surface wave magnetron, there is a possibility to design the amplifying variant of a new microwave tube. **Figure 6** presents a design of a 3-mm surface wave magnetron amplifier (amplitron). The main difference of the amplitron from magnetron lies in the fact that a anode block of the amplitron is nonresonant slow-wave structure in which an electromagnetic wave propagates from a RF input to a RF output, i.e., in an interaction space of the amplitron, there is a process exchange of electromagnetic energy between a re-entrant electron beam and traveling wave that is propagated from RF input to RF output and then to a matched load.

**Figure 7** shows the experimental dispersion characteristics of a comb-shape slow-wave structure of the 3-mm surface wave amplitron.

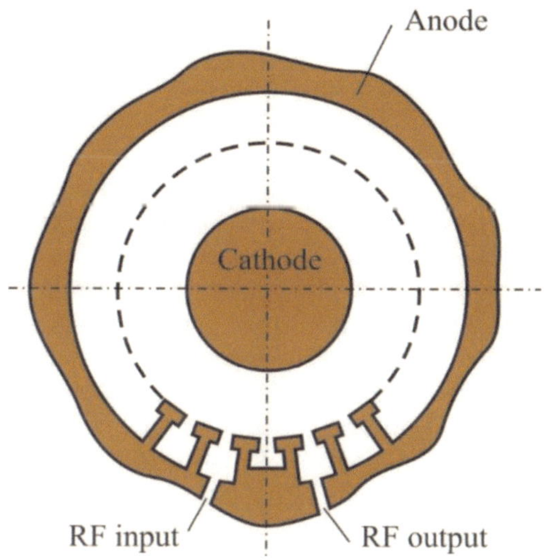


**Figure 6.**
*A scheme of interaction space of an amplitron.*

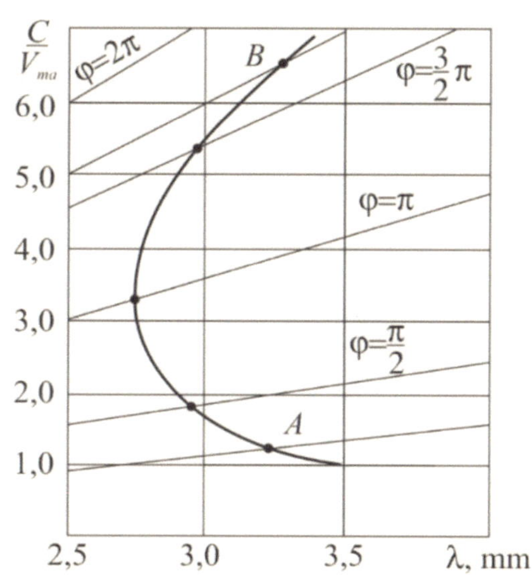


**Figure 7.**
*A scheme of interaction space of an amplitron.*

### *2.2.2 Magnetrons with two energy outputs*

Recently, there have been functional problems of different electronic systems, which became all more complex [15, 16, 18]. In particular, multifrequency radar operating in various frequency ranges for monitoring the clouds and precipitation (meteorological radar) or observing the water area and the movement of vessels in port services are increasingly applied [16]. Wherever such radars are required, there is a great quantity of interfering factors, and the multifrequency systems allow solving these problems. The operation of such radars requires a new functional element base (vacuum tubes) capable to give a functionally simple solution to a problem of multifrequency generation with high-operating characteristics. As an example of the multifrequency generator it can be a magnetron implementing the mode of an electron frequency tuning (including frequency tuning from pulse to pulse) and its application in electronic systems of different functional purpose [35, 36, 38, 39]. The practice shows [see, for example, 38, 39], that in this case in a design of the magnetron we can use two RF outputs of energy: one as active output and other a reactive one which is used for tuning a frequency.

The anode block of the magnetron with different possible variants of arrangement of the second RF output is presented in **Figure 8**. As may be seen, the second RF output of energy can be placed both symmetrically (1′) and antisymmetrically (2 и 2′) to the active RF output (1′). The main constructional and electrical parameters of the basic design of the magnetron are given in **Table 2**.

Using an existing design method of the magnetrons, described in [41], a code for computer aided design of geometry and electrical parameters of the magnetrons was developed. The computation being made with the help of this code allowed defining all parameters of the magnetron provided that a maximum current from cathode was not more than 1 A/sм$^2$.

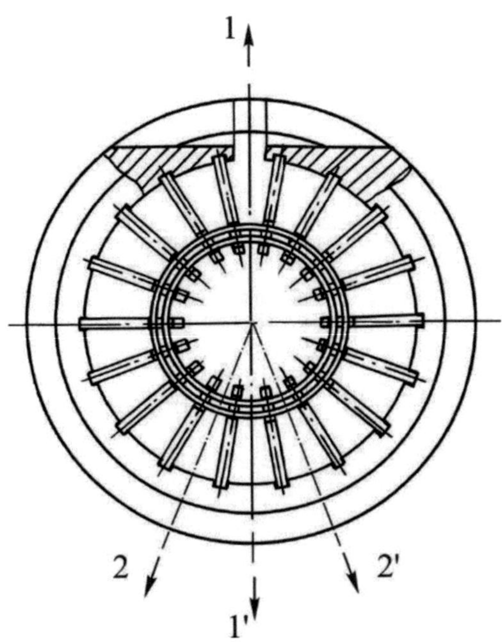


**Figure 8.**
*The possible variants of arrangement of the second RF output of energy.*

| Parameters | Symbol | Unit | Value |
|---|---|---|---|
| 1. Frequency | $f$ | GHz | 9.42 |
| 2. Cathode radius | $r_c$ | mm | 2.225 |
| 3. Anode block radius | $r_a$ | mm | 3.25 |
| 4. Anode block height | | mm | 7.0 |
| 5. Resonators number | $h$ | mm | 18 |
| 6. Anode voltage | | kV | 7.3 |

**Table 2.**
*The parameters of a magnetron.*

In order to choose the operation mode of the magnetron and to apply computer modeling using its 3-D mathematical model, it is necessary to carry out the analytical calculations to define the Hull cutoff curve

$$U = 0,022 \cdot r_a^2 \cdot B^2 \cdot \left(1 - \frac{r_c^2}{r_a^2}\right)^2$$

where $r_c$ and $r_a$ are the radiuses of the cathode and the anode in sm; $B$ is the magnetic field, Gs. Also, Hartree's voltage that can be written as

$$U_{Hartree} = U_{\min} \cdot \left[\frac{2 \cdot B}{B_0} - 1\right],$$

where

$$U_{\min} = 253 \cdot 10^3 \cdot \left[\frac{2\pi r_a}{n \cdot \lambda}\right]^2,$$

$$B_0 = \frac{21200}{n \cdot \lambda \cdot \left[1 - (r_c/r_a)^2\right]},$$

$n-$ a mode of oscillation (for $\pi-$ mode $n = \frac{N}{2}$) and $\lambda-$ wavelength in a free space.

As illustrated in **Figure 8**, as an example of the electrodynamics system of the basic design of the magnetron, we used an anode block with having the double two-sided straps. We have investigated the electrodynamics of the anode block by applying boundary conditions on an FDTD simulation.

The theoretical dependence of dispersion characteristic of the trapezoidal anode block with double two-sided straps for $\pi-$ mode $(n = \frac{N}{2})$ and three nearest to the spurious modes $\left(n = \left(\frac{N}{2} \pm 1\right), \left(\frac{N}{2} \pm 2\right), \text{and } \left(\frac{N}{2} \pm 3\right)\right)$ is shown in **Figure 9**.

As we can see from **Figure 9**, the separation between the main operative mode ($\pi-$ mode, when $n = \frac{N}{2}$) and the nearest spurious modes $\left(n = \left(\frac{N}{2} \pm 1\right)\right)$ is more than 2200 MHz. Such separation between the nearest competing modes in the magnetrons allows effectively to solve a problem of the frequency tuning in wide frequency range. On the other hand, for understanding the general situation associated with the influence of the design and axial dimensions of end regions on the shift of resonance frequency of electrodynamics system, it is necessary to carry out an additional calculation and analysis.

**Figure 10** shows a curve of shift of a resonance frequency of the anode block depending on the axial height of the end region between the vanes and end covers of electrodynamics system.

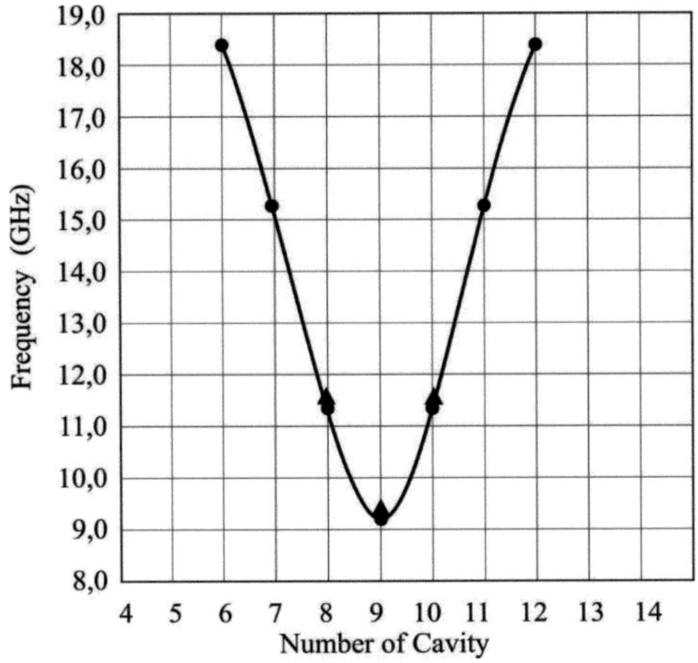


**Figure 9.**
*The results of computer modeling of the dispersion characteristic of an anode block.*

An investigation of the electrodynamics parameters (dispersion) of the anode block was carried out experimentally. A panoramic measurer of VSWR of P2–65 type was used in the measurement. By using such approach, we viewed the resonance oscillation on an operating $\pi-$ mode and nearest to the spurious modes of oscillation when $n = \left(\frac{N}{2} \pm 1\right)$. The comparison of the theoretical computation with the data of the experiment is presented in **Table 3**.

A general view of the magnetron with two energy output ports is schematically illustrated in **Figure 11**. As may be seen, the active RF output 2 of the magnetron is matched with load 5 and its reactive (passive) RF output of energy 3 is connected with a length of waveguide containing a short-circuiting piston 4. By varying the distance

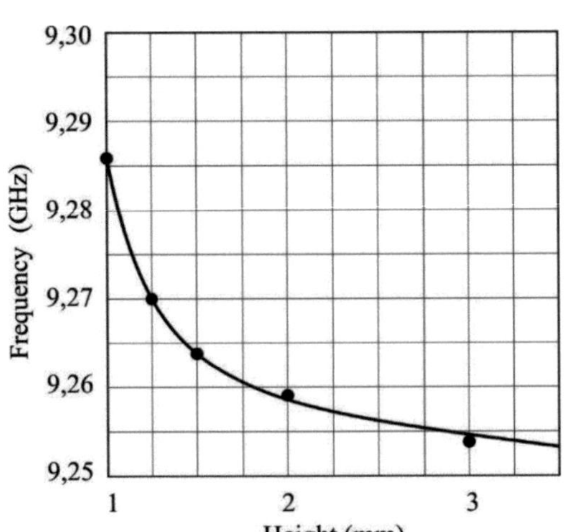


**Figure 10.**
*A resonance frequency of the "cold" anode block as a function of the distance between the vanes and end covers.*

| Parameters | Theory | Experiment |
|---|---|---|
| Frequency of operating mode $n = \frac{N}{2}$, MHz | 9253.47 | 9230.00 |
| Frequencies of nearest spurious modes $n = \left(\frac{N}{2} \pm 1\right)$, MHz | 11470.61<br>11480.56 | 11498.00<br>11828.00 |

**Table 3.**
*Comparison theory with data of experiment.*

from reactive RF output up to the short-circuiting piston, we are changing the input complex resistance of the waveguide 4 according to the following expression

$$Z_{inp} = jZ_0 \cdot tg \frac{2\pi L}{\lambda_g}, \tag{11}$$

where $Z_0$—an input characteristic impedance of a waveguide and $\lambda_g$—wavelength into waveguide. As a result, a reactive component of a complex impedance of the anode block of the magnetron is changed and a resonant frequency of the anode block is retuned.

An experimental curve of the frequency tuning for a "cold" anode block of the 3-sm magnetron with two RF outputs of energy is presented in **Figure 12**. As may be seen, changing a length of line circuit (waveguide) $L$ leads to a periodical changing a resonant frequency of the magnetron with a special period $\lambda_g/2$. As this

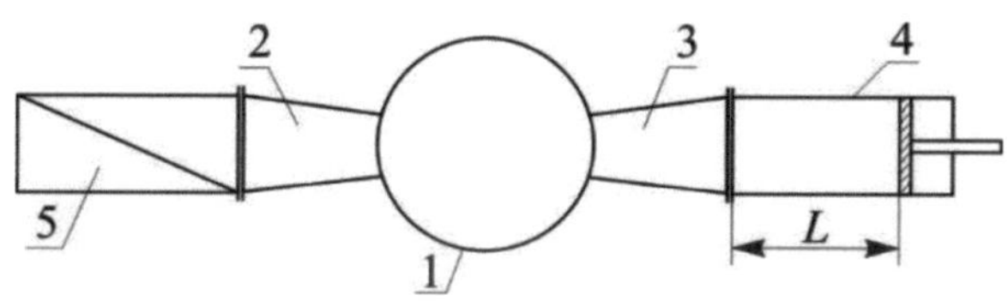


**Figure 11.**
*Schematical image of a magnetron with two RF outputs of energy. 1—an anode block of the magnetron; 2—an active RF output of energy; 3—a reactive RF output of energy; 4—a waveguide including a short-circuiting piston; 5—a matched load.*

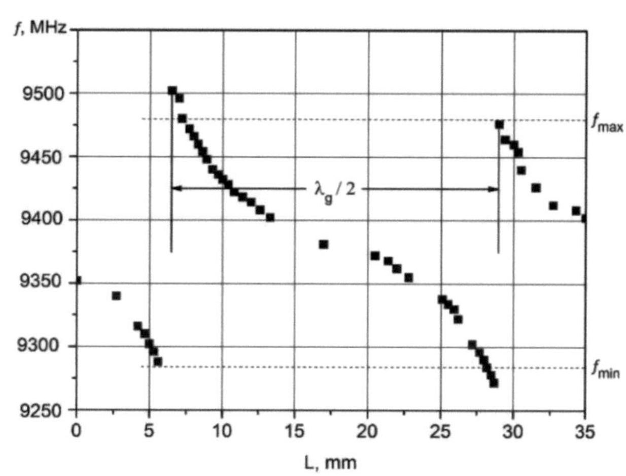


**Figure 12.**
*Experimental curve of frequency tuning in the X-band magnetron with two RF outputs.*

takes place, the full frequency tuning range is exceeded 200 MHz that is sufficient for practical application.

For comparison in **Figure 13**, we present the 3 D images and axial sections of a classical magnetron (a) and a magnetron with two RF outputs of energy (b).

### *2.2.3 Examples of application of a magnetron with two energy outputs*

The vistas of developing the magnetrons associated with stabilization of frequency and improvement of its frequency characteristics including its operation in the multifrequency mode and the electronic frequency tuning [21, 35, 36, 38].

A block diagram of a device on basis of the magnetron with two RF energy outputs and realizing a multifrequency mode of an operation with electronic tuning of a frequency from pulse to pulse is shown in **Figure 14**.

As may be seen, the given block diagram of the pulsed magnetron generator includes a magnetron with two RF outputs: active – 1 and passive – 2, as well as modulator – 4, which is synchronized with a pulsed power supply 7 for commutation of the p-i-n diodes $D_1$ and $D_2$ of an electronic switch 5. The microwave energy generated a magnetron, on the one hand, is consumed by a matched load 2 via the active RF output 1 and on the other hand is passed the reactive RF output 3 and entered on a microwave switch 5. The microwave switch is a device, which has one input and several outputs each has a load in short-circuit waveguide form length of

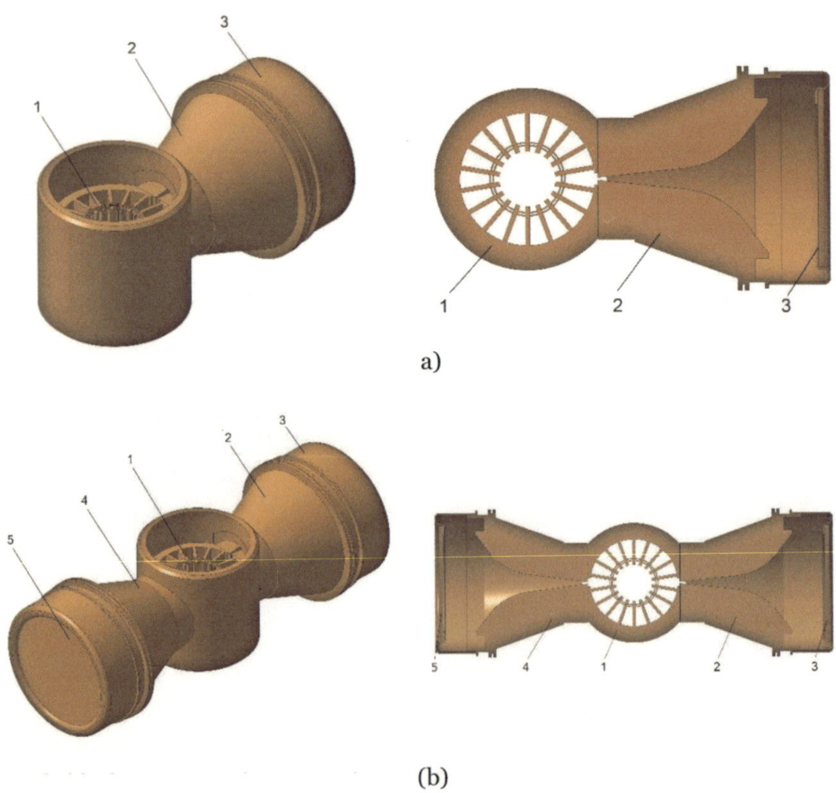


**Figure 13.**
*3D images of a classical magnetron design (a) 1—an anode block; 2—a matching transformer; 3—a RF output of energy and a magnetron with two RF output of energy (b) 1—an anode block; 2—a matching transformer of an active RF output; 3—an active RF output of energy; 4—a matching transformer of an reactive RF output; 5—an active RF outputs of energy.*

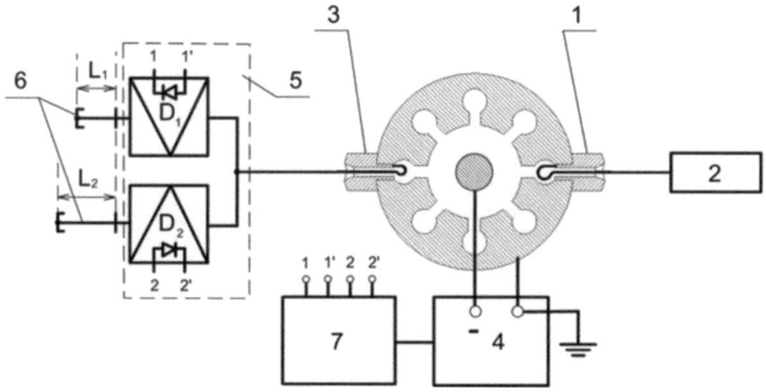


**Figure 14.**
*A block diagram of the multifrequency magnetron generator [21, 38].*

$L_i$, where $i = 1, 2, 3, \ldots N$ (in our case, $N = 2$). In the process, a value of $N$ defines a number of frequencies generated by the magnetron, generally.

In recent years, there is a major interest in the creation of the high power microwave tubes and the electronic systems on its basis [9, 15, 16, 29, 33, 42]. An analysis shows that for getting the high power microwave radiation, there are several possible approaches based on:

- an application of the relativistic microwave devices (magnetron, vircator, gyrotron, etc.);
- a generation of short and ultra-short video pulses employing the high-voltage generators (for example, Marx's generator);
- an application of nonrelativistic microwave tubes (magnetrons) and temporal resonant compression of the microwave pulses;
- a forming of focused microwave beam by application of a phased array and a system of specially distributed emitters.

The vistas of creating the high-power microwave electronics are associated with developing the high-power relativistic microwave tubes, which are provided by generation of high-power microwave pulses (peak power is units and tens GW) [9, 33]. However, the application of such devices is connected with considerable technical and technological difficulties, especially, when need to produce a compact microwave electronic system generating very short microwave pulses (duration less 100 ns). In this case for forming high power microwave pulses, there are simpler approaches based on application both the high-voltage Marx's generator and the nonrelativistic microwave tubes (magnetrons). In the last case, we supported to apply the pulse compression technology, namely the resonant microwave compression method. A central idea of this method is slow storage of electromagnetic energy in the microwave resonator and then its removal from the high factor resonator for shorter duration to a matched load (antenna) [42]. Among advantages of this method, it is necessary to note its ease of its realization, the possibility of application the industrial magnetrons, as well as the standard elements of waveguide techniques. In our case, we consider an operation of the microwave module, in which the magnetron having the two RF outputs of energy is used.

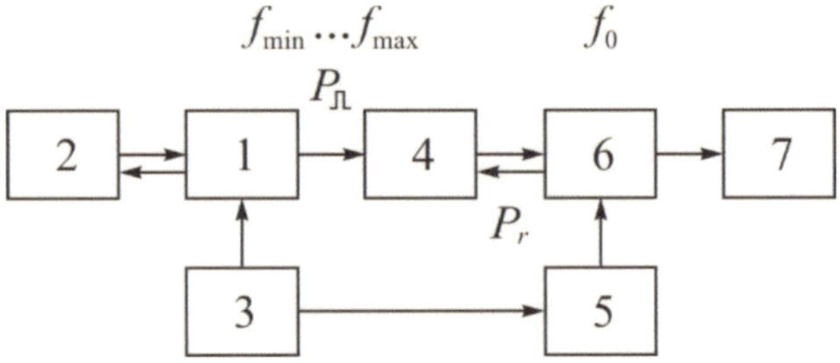


**Figure 15.**
*Block diagram of the high power microwave module [43]. 1—a magnetron with two RF outputs; 2—a tunable short-circuit waveguide; 3—a pulsed power supply (modulator); 4—ferrite isolator; 5—a generator of controlling pulses; 6—a microwave cavity; 7—a matched load (antenna).*

**Figure 15** shows a block diagram of the microwave module for temporal compression of the microwave pulses and generates the high power microwave radio pulses [43]. The magnetron that is used in this experiment has an active and a passive RF output ports. To tune the frequency of the magnetron, we used the tunable short-circuit waveguide as the reactive load of the passive RF output. The result of the frequency tuning under changes of a position $L$ of the short-circuiting piston at the reactive load of the magnetron was shown in **Figure 12**.

The analysis shows that application of the magnetron with two RF outputs in the microwave plant for forming the high power ultrashort microwave pulses allows increasing the efficiency of compression of the microwave pulses necessary to reduce a loss of power $P_r$, connected with possible reflection from microwave cavity 6 for reasons of availability of existing discrepancy between the resonant frequency of a microwave cavity $f_0$ and a frequency of the magnetron. Assuming that $f_{min} < f_0 < f_{max}$, with the help of a short-circuiting piston, we adjust an oscillation frequency of the magnetron to a resonant frequency of the microwave cavity, and as a result, the power $P_r$ reflected from microwave cavity is decreased.

In **Figure 16**, the general view of universal block diagram of the plant for generating and forming a high power microwave radiation is shown. As we notice that the radiation can be presented either as sequence of short or super-short video

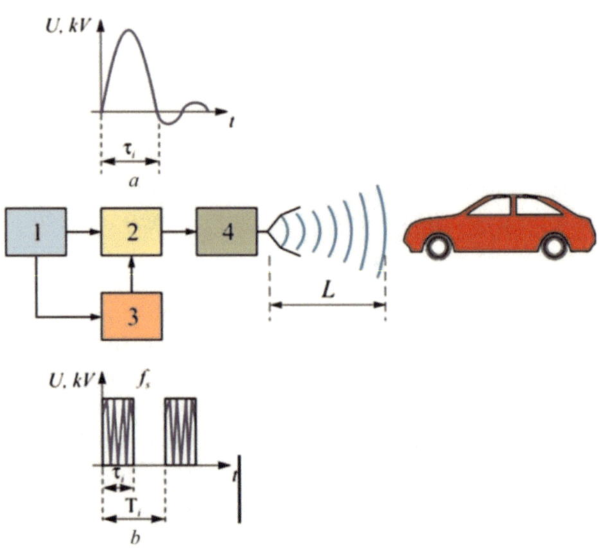


**Figure 16.**
*A block diagram of the high-power plant on basis of magnetron generator. 1—a source; 2—a microwave cavity; 3—a power supply; 4—antenna.*

pulses (case of a), or it is considered as periodical sequence of the radio-frequency pulses (case of b).

Thus, the above-mentioned examples of applying the high power generators indicate that, to date, a great demand of high power sources in many areas does not raise doubts. Also, there is a great variety of problems; the solution of which opens some new trends in the application of the high power microwave generators and the given results allow considering the employment of new designs of the mm range magnetrons and expand areas of their application in a new way.

## 3. 320 GHz TWT

### 3.1 General subjects and problems

The problem of development of the terahertz range is connected with the development of effective and compact vacuum microwave generators and amplifiers at the frequencies of 90, 220, 460, 670, 850, and 1030 GHz. TWTs are referred to as the tube, which provides the broadest band within the average power level and the most often applied in electronic systems to solve a variety of tasks in the field of communication, including space communication, probing the Earth's surface, as well as the objects of near and far space [44, 45].

The main complexity accompanying the process of producing TWT in the terahertz range consists in a contradiction between the need to combine the small size of tube interaction space (for example, the diameter of the drift channel of the delay line) and the high density of the current of an electron beam. Considering the fact that the geometrical sizes of a tube decrease in proportion to the wavelength, i.e., $D \sim \lambda$, where $D-$ is the conventional size of tube interaction space and $\lambda-$ is the wavelength in a free space, and there arise difficulties in passing an electron beam through the drift channel of the delay line. In certain cases, the given difficulties make the production of the delay line technologically impossible while applying the well-known methods and technologies. In general, these difficulties are purely technological, and they are connected not only with manufacturing the components of a tube construction (a delay line, an electron-optical system, an RF input and RF output devices, a collector, a magnetic-focusing system, etc.) but also with an assembly process of all its construction as a whole. Special consideration should be taken to the problems of providing the required critical dimensional features ($\sim$ 1–3 microns) and surface finish (roughness) of the internal surface of the delay line ($\sim$ 25–30 microns). The applied traditional technological production operations have limited opportunities already at the frequencies exceeding 400–450 GHz [46]. The specified technological difficulties have risen a considerable interest in employing programs of 3-D computer modeling of nonlinear processes of electron-wave interaction, including modeling of thermal processes in the tubes of the terahertz range. The application of the 3-D computer modeling allows defining the potentialities of the developed constructions already at the stage of tube designing in terms of ensuring the required level of power output that enables to considerably reduce the price and accelerate the development of devices as a whole.

A broad spectrum of problems accompanying the process of TWT creation in the terahertz range has led to the creation of various programs and even some new independent directions of the development of vacuum electronics in the world. First of all, it concerns a dynamically developing industry of integral vacuum microwave microelectronics, namely vacuum microelectronics of millimeter and terahertz ranges. In order to coordinate and combine the efforts of the companies which are engaged in the development of devices in the terahertz range, there have been special

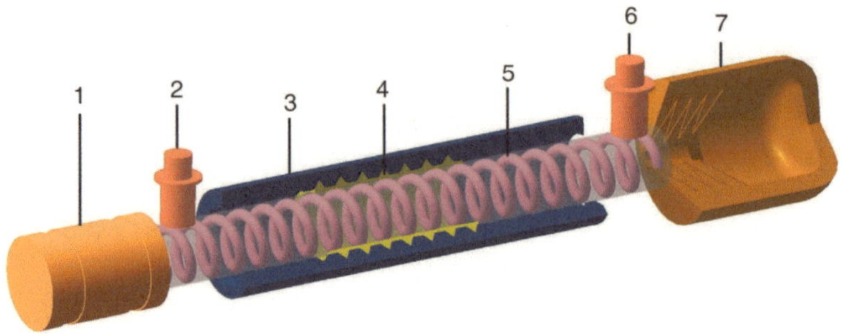


**Figure 17.**
*Scheme of the TWT design with the helix delay line. 1—an electron-optical system; 2—a RF input; 3—a magnetic focusing system; 4—a local absorber; 5—a helix-type delay line; 6—a RF output; 7—a collector.*

programs for developing terahertz technologies. In the near-term outlook, it assures a substantial progress in the field of design and development of microwave devices in a short-wave part of the terahertz range.

## 3.2 The fundamentals of TWT operation physics

The general structure scheme of classical TWT is presented in **Figure 17**. An electron beam produced by the electron-optical system 1, which includes the electron gun, accelerating and focusing electrodes, passes the delay line 5, and precipitates on the collector 7. The tube input 2 is given a RF signal, which is amplified, and output through the output 6 into matched load. In order to prevent self-excitation of the TWT in a tube between RF input and RF output, there is the energy absorber 4, the main objective of which is to reduce the wave amplitude reflected from the output 6.

One of the TWT most complex nodes is the delay line. The helix delay lines have mostly become wide spread in average power level TWT's of the centimeter and millimeter ranges. As the analysis reveals, the reduction of wavelength results in a considerable decrease in the efficiency of interaction in these systems (the value of coupling impedance decreases to a few Ohms and less), as well as some difficulties connected with the production of wire for a helix whose diameter becomes less than 50 microns. Because of this, it is necessary to have alternative types of the delay lines, which would possess acceptable electrodynamic characteristics and the ability to pass electron beams with necessary current density. In addition, these structures have to meet the demands of production simplicity while preserving thermal stability and a possibility to withdraw heat energy from its structural elements, mechanical durability, and the ability to withstand a load of various external factors (temperatures, vibrations, accelerations, etc.). It is also essential to consider that the limiting values of the dimensions of the delay lines depend on the peculiarities of their design. For this reason, the helix delay lines can be used at the frequencies, which do not exceed 60–65 GHz [47].

The analysis of publications of the last 10 years reveals that there are a number of constructions of delay lines with dimensions, which are technologically implementable in the terahertz range involving a folded waveguide, a dual comb with different types of excitation, and metal film structures on dielectric bases. In recent years, due to the extreme complexity of manufacturing in the terahertz range, classical miniature resonators and delay lines photonic crystals have been proposed to be used as delay lines [48, 49].

The existing delay lines have an essential shortcoming connected with the fact that the electromagnetic wave in these devices is surface wave, and a longitudinal z-component of an electromagnetic field, the electron beam interacting with it, decreases from the outline border of the delay line to its axis. Therefore, a great interest is the microwave plasma-filled tubes with Cherenkov's radiation mechanism in which the electromagnetic wave is volumetric [50]. As a result, efficiency of interaction process between electron beam and electromagnetic wave is raised. It leads to increasing output power in such tubes.

**Figure 18** schematically presents the design of plasma TWT. The presence of plasma allows to considerably increase the width of frequency range, to raise the output power and interaction efficiency and also enables the operational control of frequency range by implementation of tube frequency tuning both from an pulse to an pulse and within an microwave pulse. However, in order to implement the tubes for practical purposes, it is necessary to carry out additional investigations on some of unsolved problems which are related to features of beam-plasma interaction in these tubes.

### 3.3 3-D computer modeling results

As an example, let us consider the 3-D computer modeling of a delay line for TWT at the frequency of 320 GHz. As the delay line, the folded waveguide was taken, one period of which is shown in **Figure 19**. The main dimensions of the folded waveguide are presented in **Table 4**.

It is necessary to note that the folded waveguides are the most popular ones, and they are often applied to design TWT in the range up to 400 GHz due to the compact dimension, broad band, and ease in production using, for example, the UV LIGO or MEMS technologies [51–53].

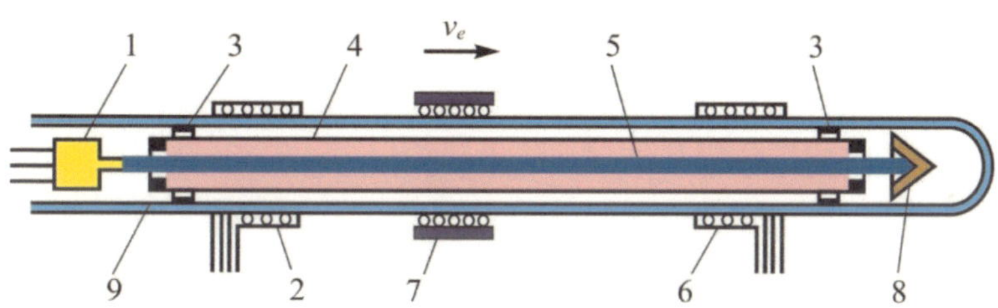


**Figure 18.**
*Scheme of plasma TWT. 1—an electron-optical system; 2—a RF input; 3—plasma ignition device; 4—plasma; 5—an electron beam; 6—a RF output; 7—an absorber; 8—a collector; 9—a bulb [50].*

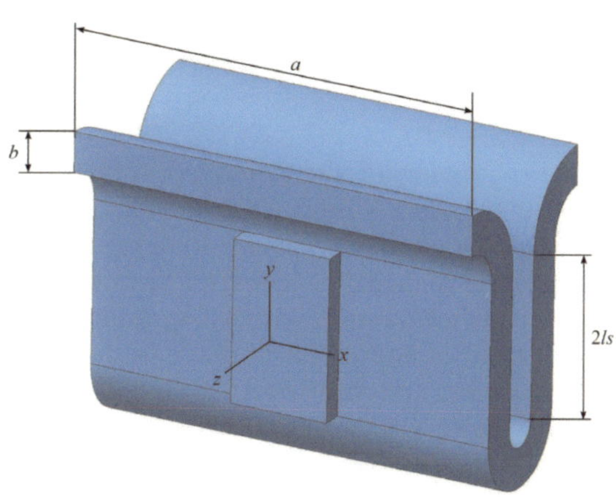


**Figure 19.**
*Illustration of one period of a folded waveguide.*

| Parameter | Symbol | Value |
|---|---|---|
| 1. Wide wall of waveguide, mm | a | 0.6 |
| 2. Narrow wall of waveguide, mm | b | 0.09 |
| 3. Pitch (one half of structure period), mm | p | 0.18 |
| 4. One half of length of straight part of waveguide, mm | ls | 0.125 |

**Table 4.**
*Parameters of a folded waveguide.*

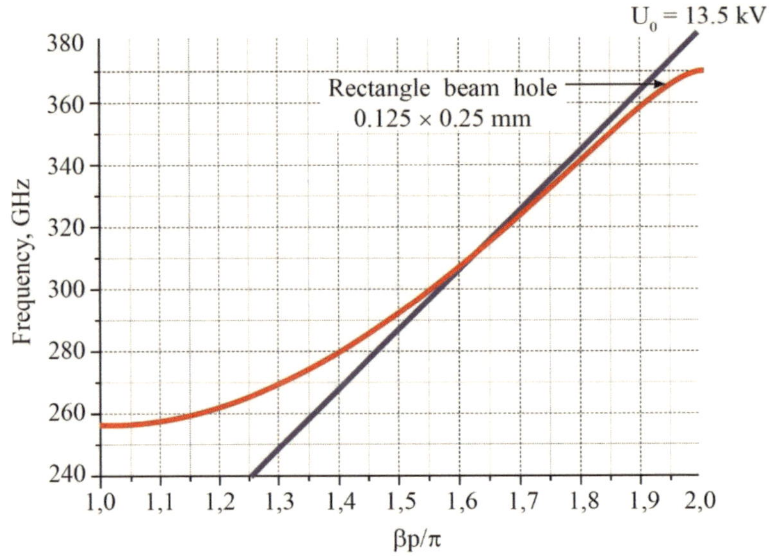


**Figure 20.**
*A dispersion diagram in case of the rectangular hole for electron beam (a red curve) and electron beam voltage (a blue curve).*

The results of computer modeling present in **Figure 20**. In this figure, a dispersion diagram and electron beam voltage are shown. As you have seen, the intersection of the curves determines an operating mode of a tube. A maximum amplification band corresponds to rectangular hole for electron beam (see **Figure 20**) whose dimensions are 0.125 × 0.25 mm. An electron beam voltage is equal 13.5 kV.

## 4. Conclusion

The current status of the theory of electron-wave interaction in a 3-mm magnetron using the mode other than $\pi-$ mode (the so-called surface-wave magnetron) and the design of the magnetron with two RF outputs are considered. It is shown that a process of phase focusing an electron beam in interaction space of the surface wave magnetron (interaction with −1 space harmonic) has a feature connected with concentration of energy of RF wave in the vicinity of a surface of the anode block. In this case for effective interaction between electron beam and RF wave, it is necessary to raise a height of electron hub of space charge in comparison with, for example, the classical magnetron. New data directed to improving the frequency characteristics of magnetrons and expanding their functionality for application in various electronic systems are obtained. In particular, the application of a pulse 2-cm magnetron with two RF outputs allows realizing a mode of electronic frequency tuning from pulse to pulse in the range 200–300 MHz. As a result of improving the frequency characteristics of magnetron generators, new circuitry was proposed for creating various electronic systems in which these magnetrons can be used.

The trend in the progress of the TWT design in the terahertz frequency range has been analyzed. Using as an example of 3-D computer modeling of a slow-wave structure as a folded waveguide, the principal possibility of designing a TWT at a frequency of 320 GHz is presented. It is shown that the advancement in the short-wave part of the terahertz range is largely associated with the search and implementation of new efficient designs of the main components of the TWT's and the slow-wave structures.

## Acknowledgements

The authors would like to thank Prof. Ye. Odarenko and Dr. V. Gerasimov for helpful discussions and numerical calculations as well as Assistance Prof. T. Frolova and Mrs. E. Isaeva for help in preparing the manuscript.

## Author details

Gennadiy Churyumov[1], Jinghui Qiu[2] and Nannan Wang[2]*

1 Kharkiv National University of Radio Electronics, Kharkiv, Ukraine

2 Harbin Institute of Technology, Harbin, China

*Address all correspondence to: wangnn@hit.edu.cn

## References

[1] Cherenkov PA. Visible emission of clean liquids by action of γ radiation. Doklady Akademii Nauk SSSR. 1934;**2**: 451. Reprinted in selected papers of soviet physicists, Usp. Fiz. Nauk 93, 385, (1967)

[2] Tamm IE, Frank IM. Coherent radiation of fast electrons in a medium. Doklady Akademii Nauk SSSR. 1937;**14**: 10 (in Russian)

[3] Jelle JV. Cerenkov radiation and its applications. British Journal of Applied Physics. 1955;**6**:277

[4] Bolotovskii BM. Vavilov-Cherenkov radiation: Its discovery and application. Physics-Uspekhi. 2009;**52**(11): 1099-1110. (in Russian)

[5] Smith SJ, Purcell HM. Visible light from localized surface charges moving across a grating. Physics Review. 1953; **92**:1069-1070

[6] Elder FR, Gurewitsch AM, Langmuir RV, Pollock HC. Radiation from electrons in a synchrotron. Physical Review. 1947;**71**(11):829-830

[7] Potylitsyn AP, Ryazanov MI, Strikhanov MN, Tishchenko AA. Diffraction radiation from relativistic particles. Springer Tracts in Modern Physics. 2011;**239**:1

[8] Ter-Mikaelyan ML. Influence of Medium on Electromagnetic Processes Under High Energies. Erevan: AS of Arm. SSR; 1969. p. 459. (in Russian)

[9] Gilmour AS Jr. Klystrons, Traveling Wave Tubes, Magnetrons, Crossed-Field Amplifiers, and Gyrotrons. Boston, MA: Artech House Inc, 2011. 883 p

[10] Proceedings of the 11-th International Vacuum Electronics Conference (IVEC 2010); 2010

[11] Dicke RH. Coherence in spontaneous radiation processes. Physics Review. 1954;**93**:99-110

[12] Hull AW. The effect of a uniform magnetic field on the motion of electrons between co-axial cylinders. Physics Review. 1921;**18**(1):13

[13] Fisk JB, Hagstrum HD, Hartman PL. The magnetron as a generator of Centimeter waves. The Bell System Technical Journal. 1946;**25**(2):167-348

[14] Collins GB. Microwave Magnetrons. New York: McGraw-Hill; 1948

[15] Military Critical Technology List. Section 8: Electronics Technology: Undersecretary of Defense, Acquisition, Technology and Logistics. USA: Pentagon; 2006. pp. 1-108

[16] Osepchuk JM. Microwave power applications. IEEE Transactions MTT. 2002;**50**(3):975-985

[17] Okress EC, editor. Microwave Power Engineering. Vol. 1, 2. New-York & London: Academic Press; 1968

[18] Edgar RH. In: Edgar RH, Osepchuk JM, editors. Consumer, Commercial, and Industrial Microwave Ovens and Heating Systems. New York: Marcel Dekker; 2001. pp. 215-278

[19] Shinohara N. Development of active phased array with phase-controlled magnetrons. In: Shinohara N, Fujiwara J, Matsumoto H, editors. International Symposium on Antennas and Propagation: Int. Conf, 21–25 Aug. Vol. 2. Fukuoka, Japan; 2000. pp. 713-716

[20] Brown WC. Satellite power system (SPS) magnetron tube assessment study. W. C. Brown. NASA Contract NAS-8-33157 for MSFC; July 10, 1980

[21] Churyumov GI, Gerasimov VP, Frolova TI, et al. The advanced designs of magnetrons with improvement output characteristics. 17th International Vacuum Electronics Conference (IVEC 2016), (19–21 April 2016 Monterey, California); 2016

[22] JSC "RPC "Istok" http://istokmw.ru/index.php?lang=en

[23] Churyumov GI, Ekezli AI. The Anomalous Increasing of the Anode Current in the Diode Structures. 17th International Vacuum Electronics Conference (IVEC 2016); (19–21 April 2016 Monterey, California); 2016

[24] JSC"Pluton"http://pluton.msk.ru/en/

[25] CPI https://www.cpii.com/division.cfm/1

[26] Nation JA et al. Advances in cold cathode physics and technology. Proceedings of the IEEE: Special Issue – New Vistas for Vacuum Electronics. 1999;**87**(5):865-889

[27] Ayzatskiy NI, Churyumov GI, Dovbnya AN, et al. Generation and formation of axially-symmetrical tubular electron beam in a cold metal secondary-emission cathode magnetron gun – Part I: Experiment. IEEE Transactions on Electron Devices. 2016; **63**(4):1704-1709

[28] Ayzatskiy NI, Churyumov GI, Dovbnya AN, et al. Generation and formation of axially-symmetrical tubular Electron beam in a cold metal secondary-emission cathode magnetron gun – Part II: Computer modelling. IEEE Transactions on Electron Devices. 2016; **63**(4):1710-1714

[29] Patent of Russia, No 2356065. The Nanosecond Radar Method Using Pulse Compression Resonance from Transmitter. Novikov SA, et al. Zayavl. 08.05.2007 (in Russian)

[30] Truten ID et al. The millimeter pulsed magnetrons using the mode of spatial harmonics. UPhJ. 1975;**20**(7): 1170-1176. (in Russian)

[31] Usikov AY, editor. Electronics and Radio Physics of Millimeter and Submillimeter Waves. Kiev: Naukova dumka; 1986. 386 p. (in Russian)

[32] Slutskin AA. The mechanism of excitation of oscillations in multi-segment magnetrons. JTPh. 1947;**XVII** (4):425-434 in Russian

[33] Magda II, Gadetski NP, Kravtsova EI, et al. Relativistic magnetron of millimeter waveband. In Proc. 18th IEEE Int. Crimean Conf. Microw. Telecommun. Technol., Sevastopol', Ukraine; 2008. pp. 637-639

[34] Churyumov GI et al. The multiperiod mathematical model of the magnetron. Radio Electronics and Informatics. 2006;**2**:15-27. (in Russian)

[35] Imran Tahir, M. Frequency and Phase Locking of a CW Magnetron (with a digital phase locked loop using pushing characteristics). Thesis. Lancaster University; September 2008. 204 p

[36] Obata H, Tsuji N, Furumoto K. Electronic-frequency-tuning magnetron. IEEE Transactions on Electron Devices. 2012;**59**(11):3111-3115

[37] Churyumov GI. The Qualitative Theory of Electron Beam Formation in a Surface Wave Magnetron (Invited Talk) 9th International Kharkiv Symposium on Physics and Engineering of Microwaves, Millimeter and Sub-Millimeter Waves (MSMW 2016), Kharkiv (Ukraine), 21-24 June, 2016. pp. 112-114

[38] Churyumov GI, Ekezli AI. Frequency tuning from pulse to pulse

magnetron generator. Patent of Ukraine # 98574, 2010 (in Ukraine)

[39] Churyumov GI, Ekezli AI. Investigation of the regime of frequency tuning in a pulsed magnetron with two outputs of energy. Elektronnaya Technika, Ser. 1, SVCh-Technika. 2014; **2**(521):39-45. (in Russian)

[40] Special Issue on New Vistas for Vacuum Electronics. Proceedings of the IEEE. 1999;**87**(5):935

[41] Shlifer ED. Calculation of Multiresonator Magnetrons. 2nd ed). M.: MEI1966. 142 p. (in Russian)

[42] Benford J, Swegle JA, Schamiloglu E. High Power Microwaves. 2nd ed. CRC Press; 2007. 556 p

[43] Churyumov GI. High-power microwave electronics: Current status, prospects of development and application features. Applied Radio Electronics. 2013;**15**(4):270-300. (in Russian)

[44] Proceedings of the 11th IEEE International Vacuum Electronics Conference (IVEC'2010); 2010

[45] Song H-J, Nagatsuma T, editors. Handbook of Terahertz Technologies Devices and Application. CRC Press; Pan Stanford Publishing; 2015. 582 p

[46] Ponomarenko SS, Kishko SA, Zavertanniy VV, et al. 400-GHz continuous-wave Clinotron oscillator. IEEE Transactions onPlasma Science. 2013;**41**(1):82-86

[47] Bushuev NA. The problem of development of the wide-band mm and sub-mm range TWTO. Journal of Radio Electronics. 2016;**11**:1-21. (ISSN 1684-1719) (in Russian)

[48] Letizia R, Mineo M, Paoloni C. Photonic crystal-structures for THz vacuum Electron devices. IEEE Transactions on Electron Devices. 2014; **62**(1):178-183

[49] Odarenko EN. Photonic crystal waveguide structures for terahertz band electronic devices. Telecommunications and Radio Engineering. 2015;**74**(3): 221-230

[50] Bernashevskiy GA, Bogdanov YB, Kislov VY, Chernov ZS. Plasma and Electronic Microwave Amplifiers and Generators. – M.: Izd-vo "Sovetskoye Radio". 1965. 95 p. (in Russian)

[51] Goldenberg BG et al., editors. Fabrication of microstructured optical elements for visible light by means of LIGA-technology. In: Nuclear Instruments & Methods in Physics Research. Sec. A. 2009;603(1/2) Proceedings of the XVII International Synchrotron Radiation conference: SR-2008, Novosibirsk, Russia, June 15–20. 2008. pp. 157-159

[52] MEMS and MOEMS Technology and Applications (SPIE Press Monograph Vol. PM85) by P. Rai-Choudhury; 2000. 516 p

[53] Tsutaki K, Neo Y, Mimura H, et al. Design on a 300 GHz band TWT with a folded waveguide fabricated by microelectromechanical systems. Journal of Infrared, Millimeter, and Terahertz Waves. 2016. DOI: 10. 1007/ s10762-016-0306-5

Chapter 3

# Design of Radial Power Combiners Based on $TE_{01}$ Circular Waveguide Mode

*José R. Montejo-Garai, Jorge A. Ruiz-Cruz and Jesús M. Rebollar*

## Abstract

Modern microwave and millimeter-wave systems require high-power amplifiers in very diverse fields such as communications or plasma physics. Although amplification technology has significantly evolved in the last decades, a single module is not enough for achieving the required power level. The solution in this case is the combination of several individual modules with power combiners. In this chapter, this concept is shown with two E-plane radial power combiners, both carrying a high-power signal with the circular waveguide $TE_{01}$ mode. The first design is a 16-way Ku-band combiner with an excellent experimental performance: return loss better than 30 dB, with a balance for the amplitudes of ±0.15 dB and ±2.5° for the phases, in a 16.7% fractional bandwidth (2 GHz centered at 12 GHz), and efficiency better than 95% in this band. The second design is a 5-way W-band combiner, showing excellent characteristics as well: the experimental prototype has a return loss better than 20 dB, with a balance for the amplitudes of ±0.4 dB and ±3.5° for the phases, in a 12.8% fractional bandwidth (12 GHz centered at 94 GHz), and efficiency better than 85% in this whole band. The experimental results obtained in both designs are the state of the art in the area of radial power combiners.

**Keywords:** radial combiner, mode transducer, mode conversion purity, *N*-way divider/combiner, evanescent mode, propagating modes, higher-order modes, return loss, isolation

## 1. Introduction

There are a large number of high-frequency systems making an extensive use of high-power modules, especially in modern systems at microwave and millimeter wave bands for research, industry and defense. Some of these applications are new communication systems with higher capacity, weather and control radars, space exploration, and scientific facilities for particle accelerators or for plasma physics [1–3]. The high-power modules required in these systems, although they may have very diverse type of specifications, require all typically high efficiency in the power amplification process, with high linearity, and over a wide frequency band. A single individual module is, in many cases, not enough to achieve this high performance in a single stage, and power combination is a classical strategy to overcome this situation. With this strategy, the requirements of the individual amplification modules are less stringent, simplifying their design, at the expense of introducing a unit for the power combination, which can be done very efficiently, as this chapter will show with two examples.

Amplification at the microwave and millimeter wave bands have been typically based on high power vacuum devices such as klystrons, magnetrons, traveling wave tubes (TWTs), etc. [4]. Although they are still a common solution at some high frequency bands, these devices may suffer from some drawbacks such as their very hard requirements of high-voltage supply, inherent thermionic noise, limited lifetime of the filament, etc. To overcome these issues implies a high cost, especially when the operation frequency increases, and some applications may not afford it.

After a continuous research over last decades, solid-state technology has become an alternative to these vacuum devices. Along with this development, power combination techniques have become crucial, for instance when a solid-state power amplifier (SSPA) module cannot provide enough power. In this case, the power combiner unit (which can be also seen, indistinctly, as a power divider because of the reciprocity for passive waveguide components) becomes a key component, which must have very stringent specifications. It must have very low insertion loss, since it has to deal with high-power signals; a high insertion loss implies less efficiency, but also high-power dissipation and heating of the unit. The different signals must be combined with a very good balance in amplitude and phase. Other important requirements are the return loss level at the input port and the isolation between the output ports.

These varieties of requirements for the power combiner have led to many topologies for implementing this function. The configuration based on a structure with a radial symmetry between the input and output ports has some intrinsic advantages in comparison with the chain-type or corporate-type combiners [5], especially when dealing with a large number of ports [6, 7]. Since the configuration has a radial symmetry, all the paths from the input to the output ports are guaranteed to be equal, providing a perfect amplitude and phase combination from the theoretical point of view (i.e., manufacturing tolerances may slightly degrade this ideal operation).

With respect to the most suitable high-frequency transmission system to implement the combiner, metallic hollow waveguides provide several advantages such as low insertion loss and high power capability. Since waveguides are larger in size in comparison to planar wave guiding structures, they exhibit higher robustness and stability, which are crucial for amplifier modules and plasma heating for fusion energy [8].

Many radial waveguide combiners can be found in the literature for Ku and Ka frequency bands. The work in [9] presents a wideband 60 GHz 16-way power divider with 20% bandwidth and 12 dB return loss level. A 19-way isolated radial combiner is presented in [10], with 12 dB return loss level in a 24.4% bandwidth at 20 GHz. A 24-way radial combiner at Ka band is proposed in [11], with 25 dB return loss level in a 15% bandwidth. A 20-way Ka-band design with 10 dB return loss level is shown in [12] with a 25% bandwidth. The design in [13] shows a very competitive performance, which will be further reviewed in this chapter.

At W-band, power combiners with radial symmetry are not as common as in lower frequency bands. A four-way design, using a corporative scheme implemented in H-plane waveguide configuration is reported in [14], with a T-junction for the division. It has a theoretical return loss (the measurement is carried out in back-to-back configuration) better than 10 dB in an 11.8% fractional bandwidth (96–108 GHz). A four-way waveguide power divider is shown in [15] with a return loss better than 13 dB in a 31.6% fractional bandwidth (80–110 GHz). This design has the output ports with 180° out of phase. A W-band solid-state power amplifier is

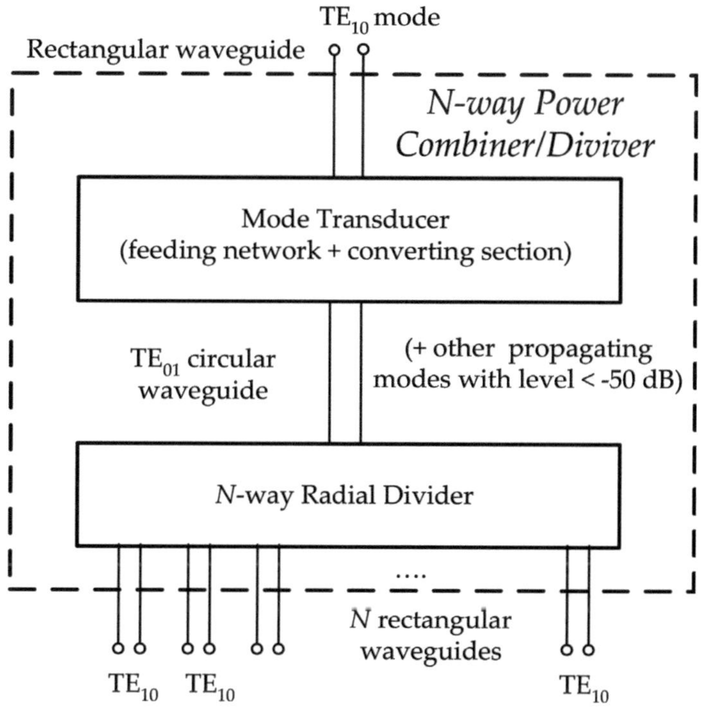


**Figure 1.**
*Scheme of the radial power combiner (or divider) made up of (i) the mode transducer between the rectangular waveguide $TE_{10}$ mode and the circular waveguide $TE_{01}$ mode, and (ii) the N-way radial divider (or combiner) dividing the power carried by the circular waveguide $TE_{01}$ mode into N rectangular waveguide $TE_{10}$ modes.*

presented in [3], using two types of waveguide combiners: a 4-way septum waveguide combiner and a 12-way radial-line waveguide combiner. The 4-way septum waveguide combiner is of the corporate type [5]. The 12-way radial-line waveguide combiner is composed of a transition from rectangular waveguide to coaxial, and then goes into the central radial-line section. Very recently [16], a 5-port W-band design has been proposed. The features of this structure are discussed in this chapter, with additional details related to the ideal S-parameters of the structure guiding the design.

When working with radial power combiners involving circular and rectangular waveguides, as the designs shown in this chapter, two main characteristics will determine the features of the final unit. The first one is the geometry of the mode transducer connected to the divider, which will be directly related to the compactness of the combiner and the level of the excited higher-order modes. The second one is the use of resistive elements or sheets within the structure to improve the isolation between the output ports [17]. In this case, the mechanical design must take into account the integration of the resistive sheets, and the insertion loss level and the power handling are usually degraded. Moreover, the amplitude and phase balance may also get worse.

Taking into account all these considerations, this chapter presents the design of two radial power combiners in waveguide technology. The prime objective has been to obtain competitive designs suitable for manufacturing at microwave and millimeter wave bands, giving priority to the following key requirements: the return loss level at the common input port, the insertion loss from the input to the different outputs, the power handling capability, and the balance of both amplitude and phase between the output ports. It will be emphasized how to control these requirements with a common strategy for designs with different number of ports and for different frequency bands. Nevertheless, for the final implementation, the different manufacturing technologies have to be taken into account in the final design process, since the control of the dimensions with respect to the tolerances and fabrication strategy (material, cuts of the parts, assembly, etc.) for the different frequency bands is crucial for achieving a successful experimental prototype.

**Figure 1** shows a detailed scheme of the topology used for the two E-plane power combiners [13, 16] discussed in this chapter. It has a mode transducer between the rectangular waveguide $TE_{10}$ mode and the circular waveguide $TE_{01}$ mode, and the $N$-way radial divider. In this structure, it will be ensured that the other propagating modes, as well as the evanescent modes, have all a level of at least 50 dB lower than the desired circular waveguide $TE_{01}$ mode just before the radial divider. This is essential to have a broadband performance and for avoiding spurious resonances in the response. The structure will be used for two designs. The first design is a 16-way Ku-band combiner, centered at 12 GHz with 2 GHz of bandwidth (16.7% fractional bandwidth). The second design is a 5-way W-band combiner, centered at 94 GHz with 12 GHz of bandwidth (12.8% fractional bandwidth).

## 2. Design of the circular waveguide $TE_{01}$ mode transducer

### 2.1 Foundation of the mode transducer operation

The radial combiner in **Figure 1** is based on the $TE_{01}$ circular waveguide mode, which is used in different fields of microwave and millimeter wave engineering. For instance, some oversized circular waveguides work with the $TE_{01}$ mode because of its low attenuation constant, since its electric field is progressively smaller when approaching the circular boundary [18–19]. Metallic cavities made up of a cylinder with circular cross section provide resonant $TE_{01p}$ modes, used for microwave filters with very low insertion loss, and this type of cavities are also common in plasma systems, gyrotrons, masers, etc. [20–23].

In all these applications, it is necessary to convert first the power coming from the generator in the fundamental mode of a suitable transmission system (typically a coaxial, or a rectangular waveguide for high frequency bands) into the $TE_{01}$ circular waveguide mode, which is not the fundamental mode of the circular waveguide. The device performing this function is the circular waveguide $TE_{01}$ mode transducer. Although there are many implementations of this device, there are two main methods for the generation of this mode from a $TE_{10}$ rectangular waveguide mode.

The first developed method was based on transforming the cross-section of the input rectangular waveguide progressively into the cross-section of the output circular waveguide, leading to a flared structure. This conversion is usually very long and involves many sections cascaded in-line following a symmetric pattern in order to prevent the generation of higher-order modes, which may degrade the overall performance [11]. Some examples are the Southworth-type converter [18, 24], the Marie-type [25, 26], and the sector converter [27]. At the beginning of these developments, the design of this type of converters was based on the expertise of the designers; nowadays, powerful tools for computer aided design (CAD) are also combined with the know-how of the designers. The main drawbacks of this kind of in-line configuration are their large length and the high level of the excited undesired modes [11].

The second method uses a sidewall coupling (by one or more several sides) between the rectangular and the circular waveguides [28]. The flower-petal transducer follows this configuration [29]. There are two main drawbacks for this structure, its narrow bandwidth and its high insertion loss level. Ka-band transducers with four branches are presented in [12, 30], showing moderate return loss level. In these references, the sidewall coupling is a simple aperture with limited degrees of freedom. As shown in [31], better return loss and wider band can be obtained by improving the sidewall coupling. Hence, the designs presented in this chapter are based on improvement made to the sidewall coupling.

The work in [31] forms the basis for the designs. **Figure 2** shows the operating principle of this type of transducer. In this figure, the converting section (dashed circle) is excited in its four sides by the $TE_{10}$ rectangular mode, generating the $TE_{01}$ mode at the circular waveguide. **Figure 2** also shows the feeding network routing the input to the four arms of the converting section.

For the design of the transducer, the classical goal is to obtain a challenging return loss at the input with high purity conversion to the circular waveguide $TE_{01}$ mode. Thus, it is essential to control the level of the non-desired modes in the circular waveguide, especially those that are propagating, with lower cutoff frequency than the $TE_{01}$ mode. A higher level for these modes degrades the conversion efficiency, but it can also lead to spurious resonances in the power divider (not always treated in detail in the literature of these devices). Thus, the first consideration in the design is to identify how the different propagating modes of the circular waveguide can be controlled, since the $TE_{01}$ mode is not its fundamental mode (i.e., it is not the mode with the lowest cutoff frequency).

This study can be done by analyzing the modes with respect to the number of symmetry planes (1, 2 or 4) of the physical structure along with the symmetry of the excitation, as in **Table 1**. **Table 1** shows the modes associated to the cases of one, two or four symmetry planes and the normalized cut-off frequencies of the modes involved in the structure. The excitation in **Figure 2** will be done with the $TE_{10}$ mode of the rectangular waveguide at the input, which has electric wall (EW) symmetry. The $TE_{01}$ mode of the circular waveguide has EW symmetry at four symmetry planes of the circular waveguide, including the planes at the four sides for the excitation. In fact, this $TE_{01}$ mode has EW symmetry for any radial plane.

The study leads to the following considerations:

a. The converting-section has four symmetry planes, all with EW boundary condition, including the radial planes crossing the sides of the excitation. Thus, according to the third column of **Table 1**, the only propagating mode in the circular waveguide under this excitation is the $TE_{01}$.

b. The feeding-network has only one physical symmetry plane.

c. Thus, the complete transducer has only one physical symmetry plane, with EW symmetry boundary condition. Therefore, in the optimization of the final transducer, the amplitudes of the propagating $TE_{11c}$ and $TE_{21c}$ modes must be kept under very low levels. The $TM_{11s}$ and $TE_{31c}$ modes have also to be controlled for avoiding higher-order mode interactions with the radial divider ($TM_{21s}$ is under cut-off in the designed bands). This will allow reducing later the length of the circular waveguide connecting the transducer and the radial divider.

## 2.2 Converting section design

The converting section has two symmetry planes with four rectangular ports at the excitation sides. This avoids the generation of the $TE_{21c}$ mode, according to the considerations in previous subsection. In addition, some kind of matching elements must be included for obtaining a challenging return loss level in broadband

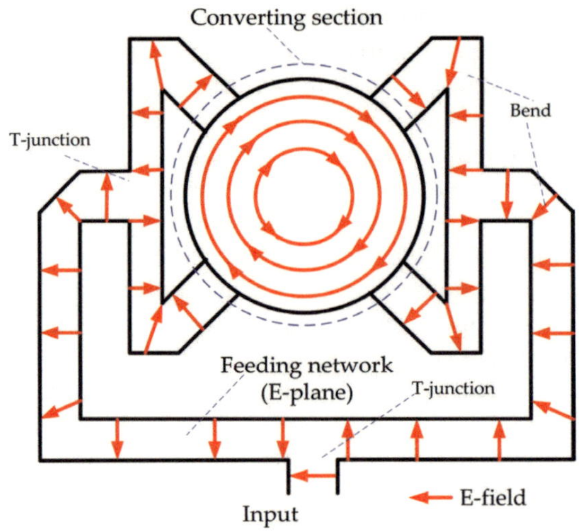


**Figure 2.**
*Scheme of a sidewall coupling for a circular waveguide $TE_{01}$ mode transducer, showing the electric field lines, the converting-section (dashed circle) excited in its four sides, and the input feeding network (with E-plane T-junctions and waveguide bends).*

**Circular waveguide modes excited under different field symmetry planes**

| **All modes** $k = \frac{f_c}{f_{cTE_{11}}}$ | **One symmetry plane electric wall (EW)** 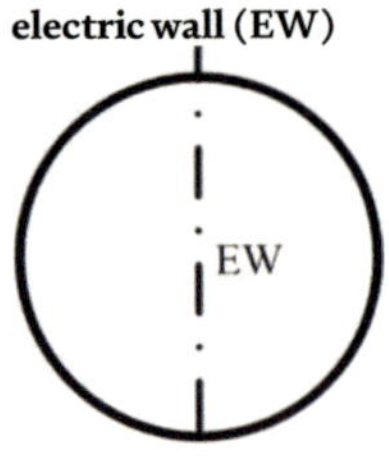  | **Two symmetry planes electric wall (EW)** 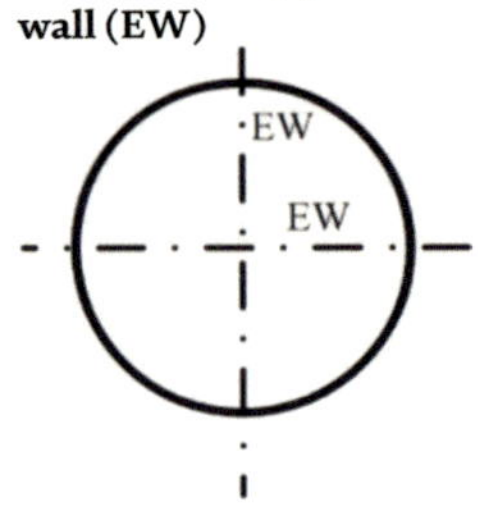  | **Four symmetry planes electric wall (EW)** 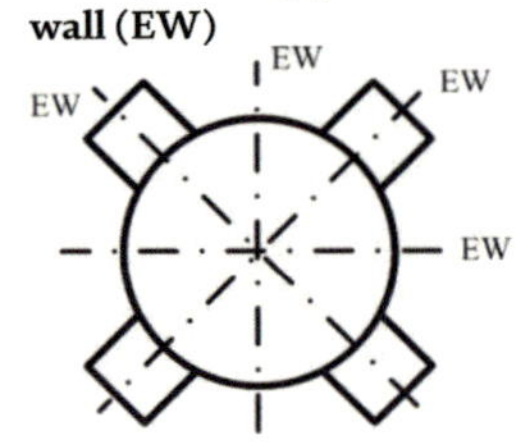  |
|---|---|---|---|
| $TE_{11c}$ k = 1 | $TE_{11c}$ | | |
| $TE_{11s}$ k = 1 | | | |
| $TM_{01}$ k = 1.31 | | | |
| $TE_{21c}$ k = 1.66 | $TE_{21c}$ | $TE_{21c}$ | |
| $TE_{21s}$ k = 1.66 | | | |
| **TE01 k = 2.08** | **TE01** | **TE01** | **TE01** |
| $TM_{11c}$ k = 2.08 | | | |
| $TM_{11s}$ k = 2.08 | $TM_{11s}$ | | |
| $TE_{31c}$ k = 2.28 | $TE_{31c}$ | | |
| $TE_{31s}$ k = 2.28 | | | |
| $TM_{21c}$ k = 2.79 | | | |
| $TM_{21s}$ k = 2.79 | $TM_{21s}$ | $TM_{21s}$ | |

*In boldface, the highlighted circular $TE_{01}$ mode is used to connect the mode transducer with the N-way radial divider.*

**Table 1.**
*Circular waveguide modes associated to the case of one, two, or four symmetry planes and their normalized cut-off frequencies.*

applications. This is done with a one-section stepped transformer shown in the insets of **Figure 3a** and **b**, connecting the circular waveguide with the rectangular waveguide ports. A circular metallic post has been also placed at the bottom of the cylinder for improving the return loss.

**Figure 3** shows the simulated response of the converting-section for both designs at Ku- and W-band, respectively, obtained with CST Microwave Studio [32]. The simulations have taken into account the four symmetry planes. In both designs, the return loss level for the $TE_{01}$ mode is better than 30 dB. In the insets of **Figure 3**, a 3D CAD view of the final converting section is included.

It is important to note that, in this structure, the only propagating mode at the circular waveguide is the $TE_{01}$, according to the third column of **Table 1**, and, thus, the reflection coefficient in **Figure 3** fully characterizes the behavior of the converting section with the considered symmetries. Moreover, it is also emphasized that the manufacturing process has been taken into account in the full-wave optimization, imposing constraints and limitations in the dimensions of the matching elements for easing the fabrication. For instance, for the W-band design that will be implemented by micromachining, corners are rounded in the simulation with 0.2 mm radius. In addition, the transformer sections keep the width of the WR10 standard waveguide used for the ports.

## 2.3 Feeding network design

The converting section is fed from the input rectangular port by means of the feeding network shown in **Figure 2**, which is composed of the following building blocks: two types of T-junctions and three types of waveguide bends, all in E-plane configuration (there is no width variation in the feeding network).

All these individual components (the building blocks), and their connection (leading to more complex building blocks), must preserve the bandwidth and the return loss level obtained previously in the converting section. Moreover, the footprint is minimized. A step-by-step process has been carried out in the design of all these individual components separately. The complete feeding network is obtained after a final optimization, varying only some connection lengths between its building blocks.

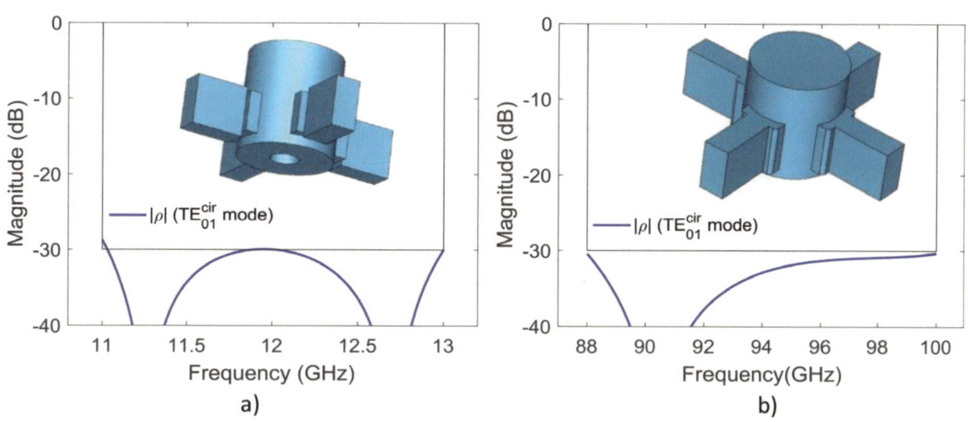


**Figure 3.**
*Simulated response of the reflection coefficient for circular waveguide $TE_{01}$ mode in the converting-section, taking into account the four symmetry planes with electric wall boundary condition (EW). In the insets, a 3D CAD view of the full converting-sections is shown. (a) Ku-band response and (b) W-band response.*

### 2.4 Final design of the transducer

The converting section and the feeding network, separately designed in the previous stages, are now connected. A final optimization is carried out in order to fulfill the specifications of the return loss level and high purity conversion from the rectangular waveguide $TE_{10}$ mode to the circular waveguide $TE_{01}$ mode. Only one-half of the converter is analyzed due to the single physical symmetry plane of the complete transducer, which has EW field symmetry. This reduces the time of the full-wave simulations. Nevertheless, since accurate results are needed in this stage, the high computational cost of the electromagnetic analyses makes crucial to minimize the number of optimization variables.

In addition, the cost function, which traditionally only involves the return loss and/or the insertion loss, must also include the level of the four higher propagating modes in the circular waveguide ($TE_{11c}$, $TE_{21c}$, $TM_{11s}$, $TE_{31c}$, according to the first column of **Table 1**), for controlling their required attenuation with respect to the desired $TE_{01}$ mode. As it could be expected, the radius of the circular waveguide is a key optimization parameter, since it controls the cutoff frequency of the modes, and it is directly related to the challenging level of 30 dB required for the return loss.

The final structure of the transducer in Ku-band is presented in the inset of **Figure 4a**, where the electric field pattern under operation is also shown. **Figure 4a** shows the simulated response achieving a return loss level higher than 30 dB, while **Figure 4b** shows the attenuation level, higher than 50 dB, for the four propagating modes in the design band from 11 to 13 GHz. A back-to-back measurement of two similar transducers manufactured in brass can be seen in [13], showing a very good agreement with respect to the theoretical simulation.

The final mode transducer for the W-band design is shown in the inset of **Figure 5a**, also with the electric field configuration. The simulated return loss, with a level better than the specified 30 dB, is shown in **Figure 5a**. The response for the attenuation is shown in **Figure 5b**, achieving levels higher than 55 dB for the four propagating modes in the design band from 88 to 100 GHz.

Both transducers have been designed in this chapter for integration with a radial divider. However, they can be also used as separated devices in diverse applications of high-energy particle accelerators or plasma heating. In all these cases, power rating is a key parameter. For the presented transducers, it has been calculated at the lowest frequency of operation in each band, i.e., 11 GHz in Ku-band and 88 GHz

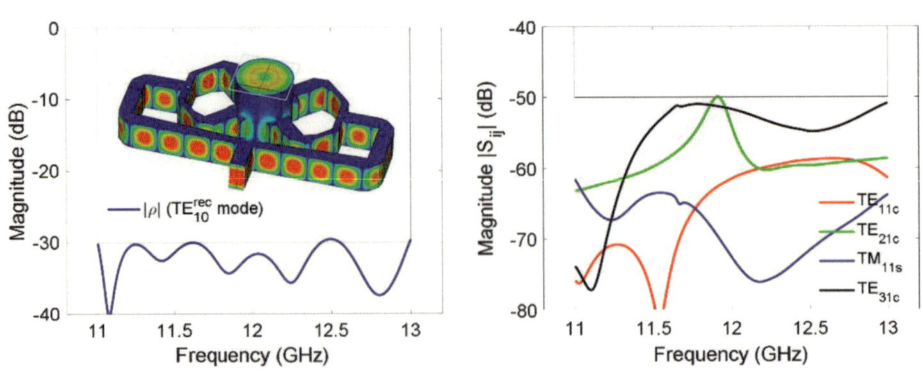


**Figure 4.**
*(a) Simulated return loss of the Ku-band transducer, better than the 30 dB goal in 2 GHz centered at f = 12 GHz (16.7%). In the inset, the electric field configuration is shown. b) Level of the four propagating modes (transducer with one physical symmetry plane having EW) at the circular waveguide (transmission from $TE^{rec}{}_{10}$), higher than 50 dB in the 11–13 GHz band.*

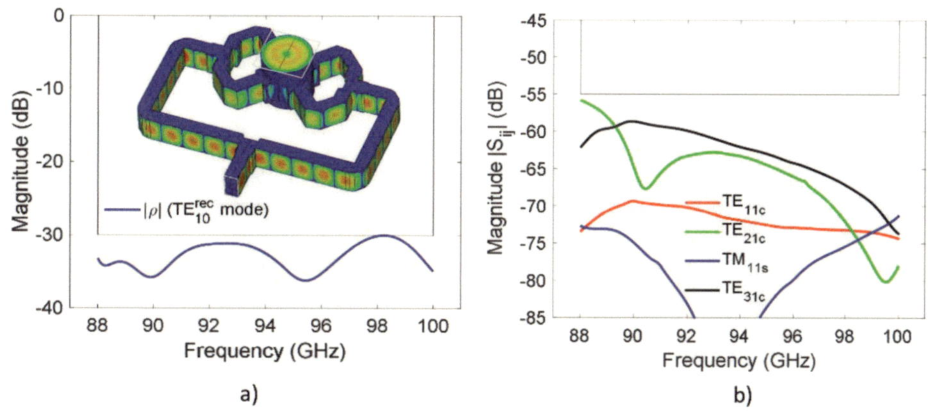


**Figure 5.**
*(a) Simulated return loss of the W-band transducer, better than the 30 dB goal in 12 GHz centered at f = 94 GHz (12.8%). In the inset, the electric field configuration is shown. b) Level of the four propagating modes (transducer with one physical symmetry plane having EW) at the circular waveguide (transmission from $TE^{rec}{}_{10}$), higher than 55 dB in the 88–100 GHz band.*

in W-band. Assuming a break down field of 30 kV/cm, and analyzing the critical dimension of each design, a 240 kW value has been obtained for the Ku-band transducer, while 9.6 kW for the W-band design.

## 3. Design of the 16-way and 5-way power dividers

The circular waveguide $TE_{01}$ mode generated by the transducer has to excite the radial divider, whose symmetry guarantees that the *N* output ports in **Figure 1** will have the same signal in amplitude and phase. Under the assumption that the radial divider will have low return loss, the power carried by the circular waveguide $TE_{01}$ mode is equally divided in magnitude and phase into the $TE_{10}$ mode of the *N* rectangular waveguides at the output ports. The radial dividers have typically a metallic post (with one or more sections) at the bottom of the structure. However, since the number *N* of output ports for the Ku- and W-band designs is very different (16 versus 5, respectively), additional strategies have been followed in each design for obtaining the specified challenging return loss level.

### 3.1 16-way radial divider

In the case of a large number of ports, as in the 16-way power divider in the Ku-band, reduced height rectangular waveguides are typically connected to the base of the radial divider [11, 12]. Since the ports are implemented in standard waveguides, in this case *N* stepped transformers (normally only changing the height) would be required between the output ports and the circular cylinder of the divider. Therefore, the complexity, the size and the insertion losses would be increased. The design shown in **Figure 6a** avoids these transformers, with standard WR75 waveguides directly attached to the divider base. Therefore, even though 16 output waveguide ports are involved in the design, the radius of the base size is not enlarged. Since the height of the waveguide is not decreased, full power handling capability is maintained, and insertion losses are not degraded. In addition, 16 transformers are avoided, simplifying the manufacturing process and reducing the cost.

The radial symmetry with appropriate EW boundary conditions is also exploited in the simulations to reduce drastically the computation time. **Figure 6b** shows the

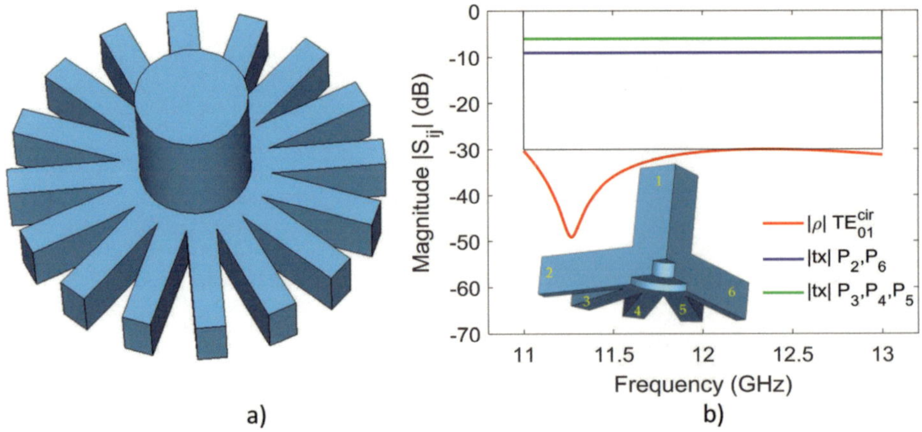


**Figure 6.**
*(a) 3D CAD view of 16-way divider with the standard WR75 waveguides directly attached to the base without stepped transformers. (b) Simulated response of the reflection coefficient of the circular waveguide $TE^{cir}{}_{01}$ mode, and the transmission to the $TE^{rec}{}_{10}$ rectangular ports (from the $TE^{cir}{}_{01}$), in the 16-way divider (only one quarter of the structure); in the inset, a 3D CAD view includes the port numbering.*

theoretical full-wave simulated response of one quarter of the divider. The return loss level is better than 30 dB in the 2 GHz bandwidth. In the same figure, it can be seen the transmission coefficients corresponding to the simulation of one quarter of the whole structure. Thus, insertion loss in ports $P_3$, $P_4$ and $P_5$ is 6 dB, but in ports $P_2$ and $P_6$, with half-height, the level is 3 dB lower than in the other three, i.e., 9 dB (see the port numbering in the inset of **Figure 6b**). After this response is obtained, the 16-way power divider is ready to be connected to the transducer.

### 3.2 5-way radial divider

The ideal S-parameter matrix of a 5-way power divider is shown in (Eq. (1)), following the notation used in [33] (**Figure 7**). The coefficients α and β represent the matching of the input and output ports, respectively. The coefficient γ is related to the input power division, while the coefficients $\eta_1$ and $\eta_2$ describe the coupling between pairs of different waveguide ports, two values in the case of $N = 5$. Assuming that the structure is lossless, the S-matrix is unitary and, consequently, it is possible to obtain the value of its elements, imposing perfect matching at the input port, i.e., α = 0. **Table 2** collects all the possible values for the other four elements (16 different matrices). It is interesting to note that in the case of $N = 5$, it is theoretically possible to match all the ports (α = β = 0).

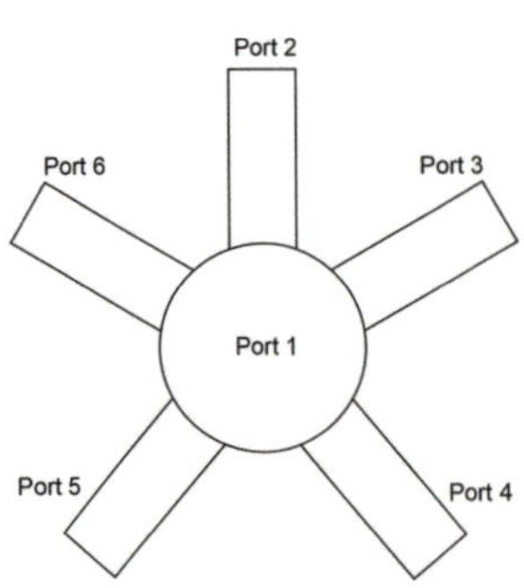


$$S = \begin{pmatrix} \alpha & \gamma & \gamma & \gamma & \gamma & \gamma \\ \gamma & \beta & \eta_2 & \eta_1 & \eta_1 & \eta_2 \\ \gamma & \eta_2 & \beta & \eta_2 & \eta_1 & \eta_1 \\ \gamma & \eta_1 & \eta_2 & \beta & \eta_2 & \eta_1 \\ \gamma & \eta_1 & \eta_1 & \eta_2 & \beta & \eta_2 \\ \gamma & \eta_2 & \eta_1 & \eta_1 & \eta_2 & \beta \end{pmatrix}$$

**Figure 7.**
*Scheme of a 5-way radial power combiner with port numbering according to (Eq. (1)).*

| $\gamma$ | $\beta$ | $\eta_1$ | $\eta_2$ | $\gamma$ | $\beta$ | $\eta_1$ | $\eta_2$ |
|---|---|---|---|---|---|---|---|
| $-\frac{\sqrt{5}}{5}$ | $\frac{4}{5}$ | $-\frac{1}{5}$ | $-\frac{1}{5}$ | $-\frac{\sqrt{5}}{5}$ | 0 | $\frac{\sqrt{5}}{5}$ | $-\frac{\sqrt{5}}{5}$ |
| $\frac{\sqrt{5}}{5}$ | $\frac{4}{5}$ | $-\frac{1}{5}$ | $-\frac{1}{5}$ | $\frac{\sqrt{5}}{5}$ | 0 | $\frac{\sqrt{5}}{5}$ | $-\frac{\sqrt{5}}{5}$ |
| $-\frac{\sqrt{5}}{5}$ | $-\frac{4}{5}$ | $\frac{1}{5}$ | $\frac{1}{5}$ | $-\frac{\sqrt{5}}{5}$ | 0 | $-\frac{\sqrt{5}}{5}$ | $\frac{\sqrt{5}}{5}$ |
| $\frac{\sqrt{5}}{5}$ | $-\frac{4}{5}$ | $\frac{1}{5}$ | $\frac{1}{5}$ | $\frac{\sqrt{5}}{5}$ | 0 | $-\frac{\sqrt{5}}{5}$ | $\frac{\sqrt{5}}{5}$ |
| $-\frac{\sqrt{5}}{5}j$ | $\frac{4}{5}$ | $-\frac{1}{5}$ | $-\frac{1}{5}$ | $-\frac{\sqrt{5}}{5}j$ | 0 | $\frac{\sqrt{5}}{5}$ | $-\frac{\sqrt{5}}{5}$ |
| $\frac{\sqrt{5}}{5}j$ | $\frac{4}{5}$ | $-\frac{1}{5}$ | $-\frac{1}{5}$ | $\frac{\sqrt{5}}{5}j$ | 0 | $\frac{\sqrt{5}}{5}$ | $-\frac{\sqrt{5}}{5}$ |
| $-\frac{\sqrt{5}}{5}j$ | $-\frac{4}{5}$ | $\frac{1}{5}$ | $\frac{1}{5}$ | $-\frac{\sqrt{5}}{5}j$ | 0 | $-\frac{\sqrt{5}}{5}$ | $\frac{\sqrt{5}}{5}$ |
| $\frac{\sqrt{5}}{5}j$ | $-\frac{4}{5}$ | $\frac{1}{5}$ | $\frac{1}{5}$ | $\frac{\sqrt{5}}{5}j$ | 0 | $-\frac{\sqrt{5}}{5}$ | $\frac{\sqrt{5}}{5}$ |

**Table 2.**
*All possible solutions for the S-matrix coefficients in an ideal 5-way radial divider.*

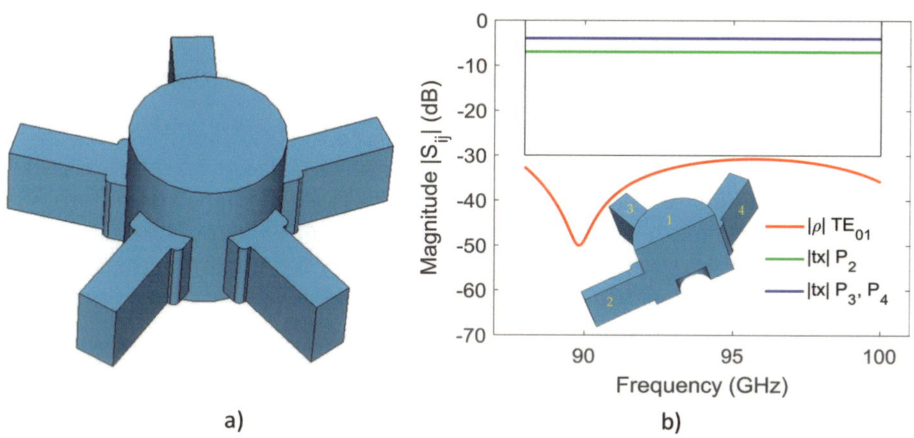


**Figure 8.**
*(a) 3D CAD view of the 5-way divider. (b) Simulated response of the reflection coefficient for the circular waveguide $TE^{cir}_{01}$ mode, and the transmission to the $TE^{rec}_{10}$ rectangular ports (from $TE^{cir}_{01}$) (only half of the structure); in the inset, a 3D CAD view including the port numbering is shown.*

For our W-band design, the 5-way power divider implementation will follow the configuration shown in **Figure 8a**. In the full-wave simulations, the radial symmetry with appropriate EW boundary conditions is exploited to reduce the computation time. **Figure 8b** shows the theoretical full-wave simulated response of the divider. The return loss level is better than 30 dB in the 12 GHz bandwidth. Since the simulation has been done over one half of the structure, the insertion loss for ports $P_3$, $P_4$ is 4.77 dB. In port $P_2$, with half-height, the level is 3 dB lower than in the others three, i.e., 7.77 dB (see the port numbering in the inset of **Figure 8b**).

## 4. Integration of the transducer with the divider: experimental results of the Ku-band and W-band power combiners

The last step in the design is the integration of the mode transducer and the radial divider. Since the return loss levels in both components have been carefully controlled, the final optimization of the full power combiner is a simple task only involving a few parameters: the radius of the circular waveguide and the dimensions of the matching cylinders in the divider and in the transducer. After that, the power combiners are manufactured and tested.

### 4.1 Experimental results of the Ku-band power combiner

The 16-way radial Ku-band power combiner has been manufactured in brass. **Figure 9a** shows a photograph of the unit during the experimental characterization of the scattering parameters. In the photograph of **Figure 9a**, the combiner has 16 high-precision WR75 matched loads attached to its output ports. The manufacturing has been done in four parts, two halves corresponding to the mode transducer and another two halves corresponding to the radial divider. The cuts separating the parts have been done along the E-plane of the waveguides, in order to reduce the insertion losses.

**Figure 9b** shows the comparison between the simulation and the measurement of the return loss level, under excitation by the input common port. In the inset, the port numbering has been included in a 3D CAD view. The agreement between theory and simulation is excellent, even when the measured value is better than 30 dB in the 11–13 GHz bandwidth. Only a small difference is shown at the upper extreme of the band at 13 GHz.

**Figure 10a** presents the measured insertion loss of the 16-way combiner compared with i) the theoretical value assuming perfect conductor ($\sigma = \infty$, −12.04 dB), and ii) the average value simulated corresponding to the brass conductivity ($\sigma = 15.9 \cdot 10^6$ S/m, −12.15 dB). From these results, it can be said that the effective conductivity obtained in the manufacturing has virtually achieved the nominal value. The amplitude balance is also very good, since the extreme values are within ±0.15 dB. The phase responses of the transmission from the input to the 16 output ports are shown in **Figure 10b**, with a detail in the inset. The balance between the extreme values is very good as well, within ±2.5°.

**Figure 11a** shows the transmission between two output ports to characterize the isolation. One of the output ports has been selected, the sixth in this case, according to the inset in **Figure 9b**, which represents a generic case because of the rotational

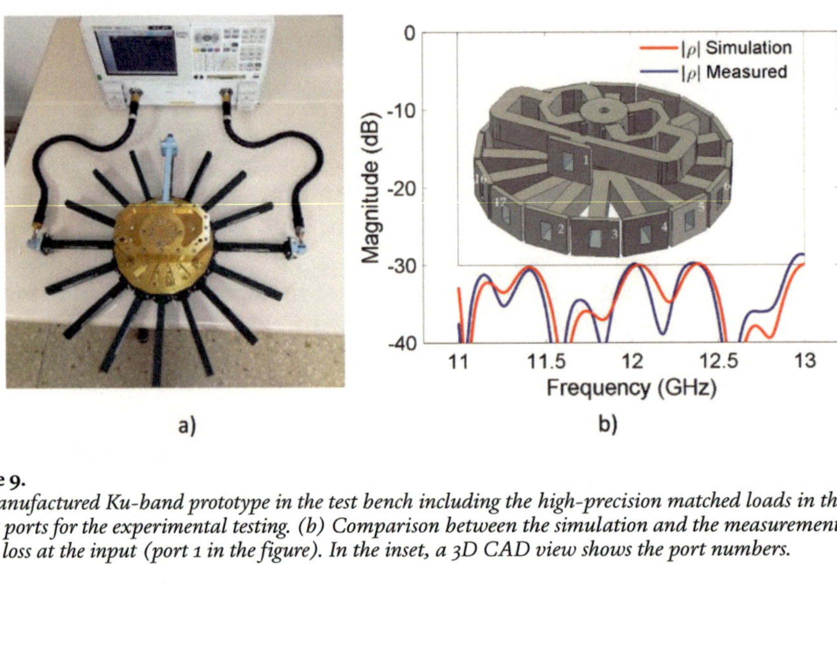


**Figure 9.**
*(a) Manufactured Ku-band prototype in the test bench including the high-precision matched loads in the output ports for the experimental testing. (b) Comparison between the simulation and the measurement of the return loss at the input (port 1 in the figure). In the inset, a 3D CAD view shows the port numbers.*

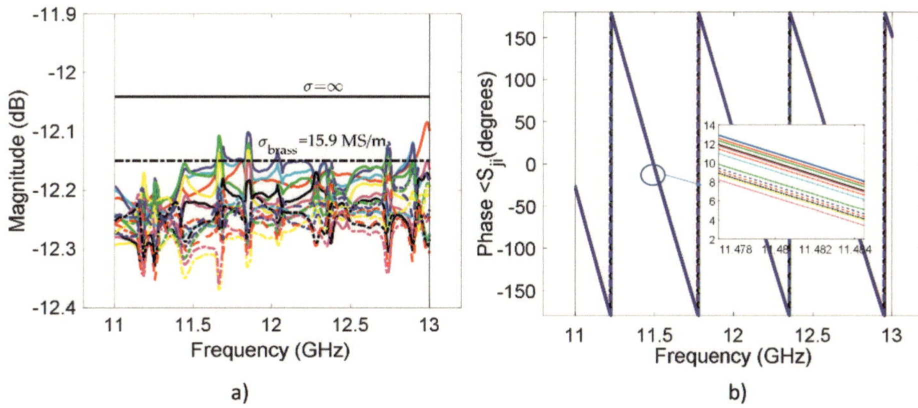


**Figure 10.**
*(a) Measured insertion loss of the 16-way Ku-band combiner compared with the theoretic value assuming perfect conductor ($\sigma = \infty$, −12.04 dB), and the average value simulated corresponding to the brass conductivity ($\sigma_{brass}$ = 15.9 MS/m, −12.15 dB). (b) Measured phases of the 16-way combiner from port 1 to ports 2–17, with a detail in the inset.*

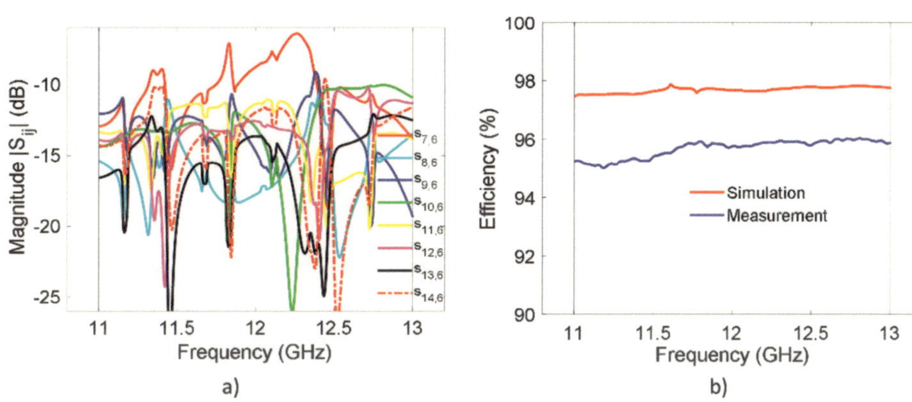


**Figure 11.**
*(a) Measured isolation response of the 16-way Ku-band power combiner: Transmission between the output port 6 (see the inset in **Figure 9b**) and the adjacent output ports of a middle of the combiner, i.e., the $S_{7,6}$, $S_{8,6}$, etc. until $S_{14,6}$ parameter. (b) Comparison between the simulated and measured combining efficiency.*

symmetry. The graph shows the measured transmission to the adjacent ports: $S_{7,6}$, $S_{8,6}$, etc., until the $S_{14,6}$ parameter. The results for the other parameters related to the other half of the divider would be similar (the simulated results would be exactly equal for a perfectly symmetric combiner). The average value for these eight responses is approximately −14 dB. However, it is interesting to note that the worst case corresponds to the isolation between two contiguous ports, the $S_{7,6}$ parameter in this case, where the minimum value is close to −6 dB.

In power combiners, a key figure of merit is the combining efficiency parameter [34], defined in (Eq. (2)), using the number 1 for the input common port. It characterizes the combined effect of the deviations of both magnitude and phase with respect to the ideal behavior. **Figure 11b** shows the simulated and measured efficiency, which is better than 95% in the whole operating bandwidth.

$$\xi = \frac{1}{N}\left|\sum_{k=1}^{N} S_{k+1,1}\right|^2 \qquad (1)$$

### 4.2 Experimental results of the W-band power combiner

The 5-way radial W-band power combiner has been also manufactured in brass [16] by micromachining. The unit has been divided into four parts, which will be stacked vertically. **Figure 12a** shows the CAD view of the parts separated before the assembly, and also after the integration. **Figure 12b** shows a photograph of the combiner during the experimental characterization of the scattering parameters, with high-precision WR10 matched loads.

The comparison between the simulation and the measurement of the return loss level is shown in **Figure 13a**. This is the reflection coefficient seen at the input common port, which is port 6 in the numbering in the inset of the figure. The measured level is better than 20 dB in the complete operation band (12 GHz centered at 94 GHz), and better than 25 dB in the 80% of this bandwidth. These measured levels are coherent with the sensitivity analysis of the power divider taking into account a tolerance for the fabrication of ±0.02 mm.

A systematic process has been followed to characterize the insertion loss of all the transmissions between the common input and the five outputs. According to

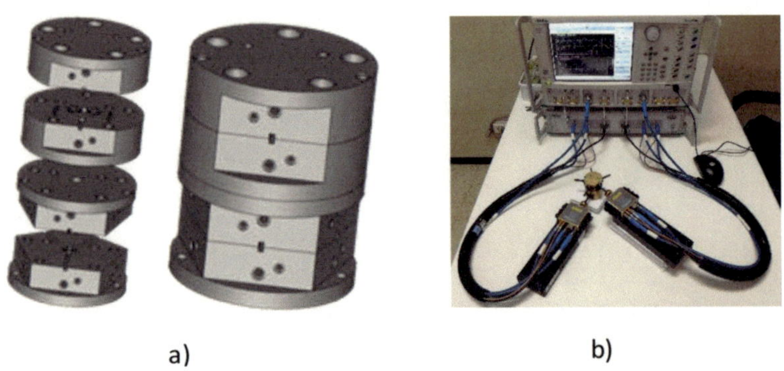


**Figure 12.**
*(a) 3D CAD of the four sections in E-plane configuration to make the 5-way W-band combiner and its final assembly. (b) Manufactured prototype in the measurement bench including the high-precision matched loads in the output ports for the experimental characterization.*

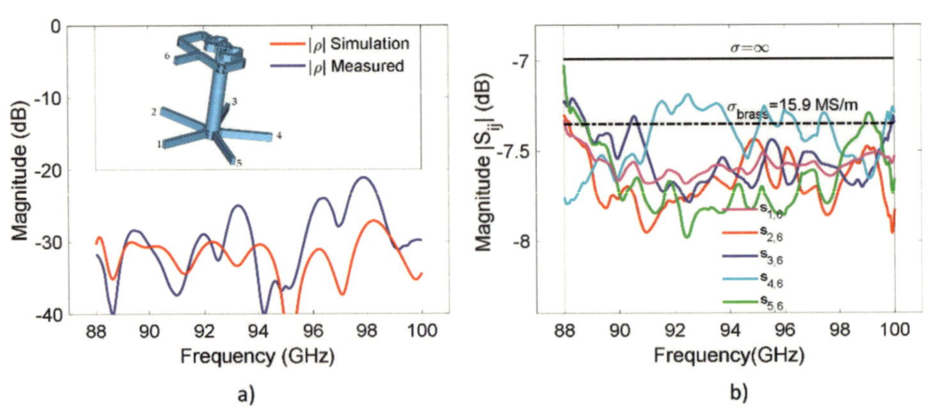


**Figure 13.**
*(a) Comparison between the simulation and the measurement corresponding to the return loss level of the 5-way power combiner excited by the input common port (port 6 in the figure). In the inset a 3D CAD view is shown including the port numbers. (b) Measured insertion loss of the 5-way combiner compared with the theoretical value assuming perfect conductor ($\sigma = \infty$, −6.99 dB), and the average value simulated corresponding to the brass conductivity ($\sigma_{brass}$ = 15.9 MS/m, −7.35 dB).*

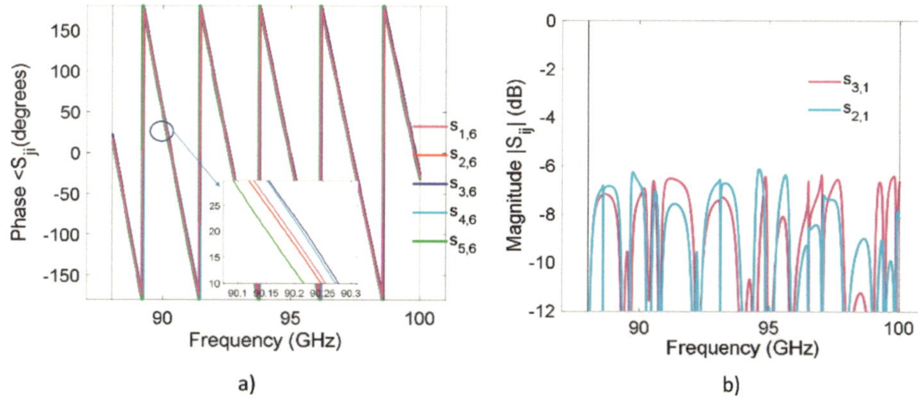


**Figure 14.**
*(a) Measured phases of the 5-way combiner, with a detail in the inset showing its very small difference. (b) Measured isolation response corresponding to the transmission between output port 1 (see the inset in* **Figure 13a***), and the adjacent output ports of a middle of the combiner, i.e., the $S_{2,1}$ and $S_{3,1}$, parameters.*

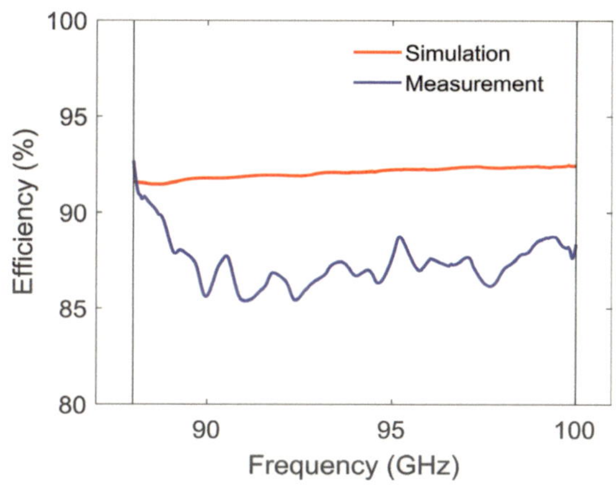


**Figure 15.**
*Comparison between the simulated and measured combining efficiency of the 5-way W-band power radial combiner.*

the numbering in the inset of **Figure 13a**, the vector network analyzer is connected between port 6 and the corresponding port from 1 to 5, while the other four ports are connected to high precision matched loads (i.e., with return loss level better than 40 dB at W-band).

**Figure 13b** presents the measured insertion loss of the 5-way combiner compared with the theoretical value assuming perfect conductor ($\sigma = \infty$, −6.99 dB), and the average value simulated corresponding to the brass conductivity ($\sigma_{brass} = 15.9{\cdot}10^6$ S/m, −7.35 dB). The average measured value of −7.6 dB implies that the effective conductivity obtained in the manufacturing is only slightly degraded with respect to the nominal value. The balance for the amplitudes is very good, since it is within ±0.4 dB at the extremes of the band.

The phase response of the transmission between the input and the five output ports is shown in **Figure 14a**, with a difference at the extremes of the band within ±3.5°, which also emphasizes the accurate manufacturing for obtaining this reduced margin at this band. **Figure 14b** shows the transmission between two output ports to characterize the isolation. Since the structure is symmetric, a generic port is selected (number 1 in this case, using the numbering in the inset of **Figure 13a**). In consequence, it is simulated and measured the transmission to its adjacent ports, i.e., the $S_{2,1}$ and $S_{3,1}$ parameter. Their average level is approximately −7 dB, very close to the theoretical value of $|\eta_1| = |\eta_2| = |\sqrt{5}/5|$ in **Table 2**, which is −6.98 dB. Finally, the efficiency has been simulated and measured showing both results in **Figure 15**. The measured efficiency is better than 85% in the whole operating bandwidth.

## 5. Conclusions

Two radial power combiners based on the $TE_{01}$ mode of the circular waveguide have been designed, manufactured, and tested. They can be used in high-frequency systems for diverse applications at microwave and millimeter wave bands. A systematic step-by step process has been followed to control the partial responses of the two main building blocks of the combiner, i.e., the mode transducer and the radial divider. In all the steps, the symmetry of each building block has been taken into account to reduce the computational effort in the optimization process. Moreover, the geometries have been simplified as much as possible to ease the manufacturing and to reduce the cost.

For the mode transducers, a strict control of the modes in the problem has been done by means of a rigorous analysis of the symmetry planes and the mode families involved in the structures. The sidewall excitation has led to very compact structures, which can also be used for mode transduction in other fields such as plasma heating.

In the case of the Ku-band 16-way divider, the output waveguides are directly connected to the divider without increasing the radius of the cylinder, avoiding extra stepped transformers. This improves the power handling and simplifies the manufacturing with respect to other designs using reduced height waveguides. In the case of the W-band 5-way divider, a careful design and the manufacturing by means of high precision micromachining with stacked waveguides have led to an excellent experimental performance.

The experimental results in both Ku- and W-bands are, to the authors' knowledge, the state-of-art of radial combiners based on the $TE_{01}$ mode of the circular waveguide. This statement is based on the obtained results in return and insertion loss, and balance for both the phase and the amplitude. The Ku-band 16-way divider power combiner has achieved a 95% measured efficiency in a 16.7% fractional bandwidth. The W-band 5-way divider power combiner shows an 85% measured efficiency in a 12.8% fractional bandwidth.

## Acknowledgements

This work was supported by the Spanish government (Agencia Estatal de Investigación) under grants (ADDMATE) TEC2016-76070-C3-1/2-R (AEI/FEDER/UE) and the program of Comunidad de Madrid S2013/ICE-3000 (SPADERADARCM).

The authors would like to thank INMEPRE S.A., the diligence in the manufacturing.

## Author details

José R. Montejo-Garai[1*], Jorge A. Ruiz-Cruz[2] and Jesús M. Rebollar[1]

1 Grupo de Electromagnetismo Aplicado, Information Processing and Telecommunications Center, Universidad Politécnica de Madrid, Madrid, Spain

2 Escuela Politécnica Superior, Universidad Autónoma de Madrid, Madrid, Spain

*Address all correspondence to: jr@etc.upm.es

## References


[1] Khan P, Epp L, Silva A. A Ka-band wide-bandgap solid-state power amplifier: Architecture performance estimates. IPN Progress Report 42-164. 2005;**42**:163

[2] Kanto K, Satomi A, Asahi Y, Kashiwabara Y, Matsushita K, Takagi K. An X-band 250W solid-state power amplifier using gain power HEMTS. In: 2008 IEEE Radio and Wireless Symposium. 2008. pp. 77-80

[3] Schellenberg J, Watkins E, Micovic M, Kim B, Han K. W-band, 5W solid-state power amplifier/combiner. In: 2010 IEEE MTT-S International Microwave Symposium Digest (MTT). 2010. pp. 240-243

[4] Qiu JX, Levush B, Pasour J, Katz A, Armstrong CM, Whaley DR, et al. Vacuum tube amplifiers. IEEE Microwave Magazine. 2009;**10**(7):38-51

[5] Epp LW, Hoppe DJ, Khan AR, Stride SL. A high power Ka-band (31-36 GHz) solid-state amplifier based on low-loss corporate waveguide combining. IEEE Transactions on Microwave Theory and Techniques. 2008;**56**(8):1899-1908

[6] Russell KJ. Microwave power combining techniques. IEEE Transactions on Microwave Theory and Techniques. 1979;**27**(5):472-478

[7] Chang K, Sun C. Millimeter-wave power-combining techniques. IEEE Transactions on Microwave Theory and Techniques. 1983;**31**:91-107

[8] Montejo-Garai JR, Leal-Sevillano CA, Ruiz-Cruz JA, Rebollar JM, Estrada T. Synthesis and design of waveguide band-stop filters without out-of-band spurious responses for plasma diagnosis. Fusion Engineering and Design. 2012;**89**(9):1662-1666

[9] Hsu T-I, Simonutti MD. A wideband 60 GHz 16-way power divider/combiner network. IEEE MTT-S International Microwave Symposium Digest. 1984, 1984;**1984**:175-177

[10] Chen M. A 19-way isolated power divider via the $TE_{01}$ circular waveguide mode transition. In: 1986 IEEE MTT-S International Microwave Symposium Digest. 1986. pp. 511-513

[11] Epp L, Khan P, Silva A. Ka-band wide-bandgap solid-state power amplifier: Hardware validation. IPN Progress Report. 2005;**42-163**

[12] Chu Q-X, Mo D-Y, Wu Q-S. An isolated radial power divider via circular waveguide-mode transducer. IEEE Transactions on Microwave Theory and Techniques. 2015;**63**(12):3988-3996

[13] Montejo-Garai JR, Saracho-Pantoja IO, Ruiz-Cruz JA, Rebollar JM. High-performance 16-way Ku-band radial power combiner based on the TE01-circular waveguide mode. Review of Scientific Instruments. 2018;**89**:034703

[14] Zhang F, Song K, Li G, Zhao M. Sub-THz four-way waveguide power combiner with low insertion loss. International Journal of Infrared and Millimeter Waves. 2014;**35**:451-457

[15] Zhang F, Song K, Fan M, Fan Y. All-metal-waveguide power divider with high power-combining efficiency. International Journal of Infrared and Millimeter Waves. 2016;**37**:258-266

[16] Montejo-Garai JR, Ruiz-Cruz JA, Rebollar JM. 5-way radial power combiner at W-band by stacked waveguide micromachining. Nuclear Instruments and Methods in Physics Research Section A: Accelerators, Spectrometers, Detectors and Associated Equipment. 2018;**905**:91-95

[17] Khan P, Epp L. Ka-band wide-bandgap solid-state power amplifier:

Prototype combiner spurious mode suppression and power constraints. IPN Progress Report. 2006;**42-163**

[18] Southworth G. Principles and Applications of Waveguide Transmission. The Bell Telephone Laboratories Series. Van Nostrand; 1956

[19] Leal-Sevillano CA, Ruiz-Cruz JA, Montejo-Garai JR, Rebollar JM. Field propagation in circular hollow waveguides with non-ideal metallic conductors from microwaves to terahertz frequencies. IEEE Transactions on Microwave Theory and Techniques. 2011;**59**(12):3013-3022

[20] Chu K. The electron cyclotron maser. Reviews of Modern Physics. 2004;**76**(2):489

[21] Blank M, Danly BG, Levush B. Experimental demonstration of a W-band (94 GHz) gyrotron amplifier. IEEE Transactions on Plasma Science. 1999;**27**(2):405-411

[22] Leou K, McDermott D, Balkcum A, Luhmann N. Stable high-power $TE_{01}$ gyro-twt amplifiers. IEEE Transactions on Plasma Science. 1994;**22**(5):585-592

[23] Garven M, Calame JP, Danly BG, Nguyen KT, Levush B, Wood F, et al. A gyrotron-traveling-wave tube amplifier experiment with a ceramic loaded interaction region. IEEE Transactions on Plasma Science. 2002;**30**(3):885-893

[24] Wolfert P. A wide-band rectangular-to-circular mode transducer for millimetre waves (correspondence). IEEE Transactions on Microwave Theory and Techniques. 1963;**11**(5):430-431

[25] Marie G. British Parent 805519; Dec 1958

[26] Saad S, Davies J, Davies O. Analysis and design of a circular $TE_{01}$ mode transducer. IEE Journal on Microwaves, Optics and Acoustics. 1977;**1**(2):58-62

[27] Xue Q-Z, Zhang S-C, Liu P-K. Design of the $H_{10}$ to $H_{01}$ sector-type mode converter at Ka-band. International Journal of Infrared and Millimeter Waves. 2005;**26**(10):1407-1415

[28] Nusinovich G, Barker R, Luhmann N, Booske J. Modern microwave and millimetre-wave power electronics. 2005. Chapter 11

[29] Hoag H, Tantawi S, Callin R, Deruyter H, Farkas Z, Ko K, et al. Flower-petal mode converter for NLC. In: Particle Accelerator Conference Proceedings of the IEEE. 1993. pp. 1121-1123

[30] Yu C-F, Chang T-H. High-performance circular $TE_{01}$-mode converter. IEEE Transactions on Microwave Theory and Techniques. 2005;**53**(12):3794-3798

[31] Montejo-Garai JR, Saracho-Pantoja IO, Ruiz-Cruz JA, Rebollar JM. Broadband and high-purity Ku-band circular $TE_{01}$-mode converter. In: 2016 Asia-Pacific Microwave Conference (APMC). 2016. pp. 1-4

[32] CST, Computer Simulation Technology. Available from: https://www.cst.com/

[33] Holzman EL. An eigenvalue equation analysis of a symmetrical coax line to N-way waveguide power divider. IEEE Transactions on Microwave Theory and Techniques. 1994;**42**(I):1162-1166

[34] Gupta MS. Power combining efficiency and its optimisation. Proceedings of the Institute of Electrical and Electronics Engineers: Microwaves, Antennas & Propagation. 1992;**139**(3, pt. H):233-238

Section 3

# Electromagnetic Waves Interaction in Different Media

Chapter 4

# Energy Transfer from Electromagnetic Fields to Materials

*Graham Brodie*

## Abstract

Electromagnetic fields are complex phenomena, which transport energy and information across space. Information can be imposed onto electromagnetic waves by human ingenuity, through various forms of modulation; however, this chapter will focus on the acquisition of information as electromagnetic waves are generated by materials or pass through materials. The chapter will also consider how energy is transferred to materials by electromagnetic fields.

**Keywords:** electromagnetic propagation, dielectric properties, information acquisition, dielectric heating

## 1. Introduction

Electromagnetic fields are a complex phenomenon because they can propagate through vacuum without the need for a material medium, they simultaneously behave like waves and like particles [1, 2], and they are intrinsically linked to the behaviour of the space–time continuum [3]. It can be shown that magnetic fields appear through relativistic motion of electric fields, which is why electricity and magnetism are so closely linked [4]. It has even been suggested that electromagnetic phenomena may be a space–time phenomenon, with gravitation being the result of space–time curvature [3] and electro-magnetic behaviour being the result of space–time torsion [5].

James Clerk Maxwell developed a theory to explain electromagnetic waves. He summarised this relationship between electricity and magnetism into what are now referred to as “Maxwell’s Equations.” An EM wave is described in terms of its:

1. Frequency (f), which is given the unit of Hertz (Hz);

2. Wavelength ($\lambda$), which is the distance between successive crests or troughs in the wave (m); and

3. Speed ($c$), which is measured in metres per second.

These three properties are related by the equation:

$$c = \lambda f \tag{1}$$

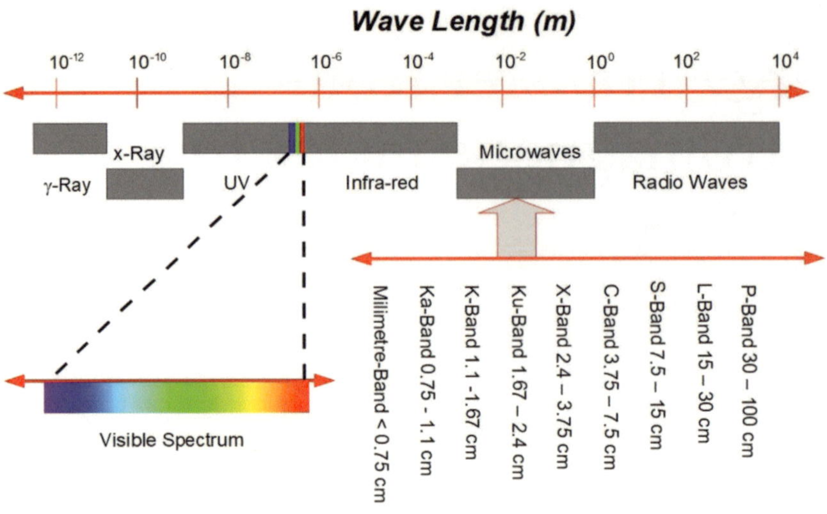


**Figure 1.**
*Schematic of the electromagnetic spectrum.*

The speed of the electromagnetic wave is determined by:

$$c = \frac{1}{\sqrt{\mu\varepsilon}} \tag{2}$$

where ε is the electrical permittivity of the space in which wave exists and μ is the magnetic permeability of the space in which the wave exists.

Electromagnetic waves can be of any frequency; therefore, the full range of possible frequencies is referred to as the electromagnetic spectrum. The known electromagnetic spectrum extends from frequencies of around $f = 3 \times 10^3$ Hz ($\lambda = 100$ km) to $f = 3 \times 10^{26}$ Hz ($\lambda = 10^{-18}$ m), which covers everything from ultra-long radio waves to high-energy gamma rays [6]. A schematic of the electromagnetic spectrum is shown in **Figure 1**.

Electromagnetic waves can be harnessed to: transmit information; acquire information from a medium; or transmit energy. The first category of applications includes: terrestrial and satellite communication links; the global positioning system (GPS); mobile telephony; and so on [7]. The second category of applications includes: radar; radio-astronomy; microwave thermography; remote sensing and detection, and material property measurements [8]. The third category of applications is associated with electromagnetic heating and wireless power transmission.

This chapter will focus on the interactions of electromagnetic waves with materials and will therefore include acquiring information from a medium and transition of energy.

## 2. Some background theory

When electromagnetic waves encounter materials, the wave will be partially reflected, attenuated, delayed compared with a wave travelling through free space [9–11], and repolarised. Surface interactions, such as reflection, refraction, transmission and repolarisation reveal important information about the material and its immediate environment. For example, **Figure 2** shows how reflection from the surface of a pond and transmission of light from in the pond water reveal

**Figure 2.**
*The combination of surface reflection and transmission through the water in this pond reveal the fish in the water and the trees in the immediate environment of the pond, hence visible light can acquire and transfer this information through space.*

information about what is in and around the pond. All interactions between electromagnetic waves and materials are governed by the dielectric properties of the material and how these properties alter the electrical and magnetic properties of the space occupied by the material.

### 2.1 Dielectric properties of materials

All materials alter the space, which they occupy. Because materials are composed of various charged particles, these alterations include changes to the electrical and magnetic behaviour of space. These properties are described by the electrical permittivity and magnetic permeability of the space, which the electromagnetic fields encounter.

Magnetic permeability is the measure of the ability of a material to support the formation of a magnetic field within itself. The magnetic permeability of space is: $\mu_o = 4\pi \times 10^{-7}$ (H/m). Except in the case of ferromagnetic materials, the magnetic permeability of many materials is equivalent to that of free space. The magnetic permeability of ferromagnetic materials varies greatly with field strength.

Space itself has dielectric properties [12] with an electrical permittivity of approximately $\varepsilon_o = 8.8541878 \times 10^{-12}$ or $\varepsilon_o \approx \frac{1}{36\pi} \times 10^{-9}$ F m$^{-1}$. Electrical permittivity describes the amount of charge needed to generate one unit of electric flux in a medium. All materials increase the electrical permittivity of the space they occupy, compared with free space (vacuum); therefore, some materials can support higher electric flux than free space. These materials are referred to as dielectric materials.

Debye [13] studied the behaviour of solutions with polar molecules and consequently refined the complex dielectric constant, which includes the conductivity of the material. His final equation became:

$$\varepsilon = \varepsilon_\infty + \frac{\varepsilon_s - \varepsilon_\infty}{1 + j\omega\tau} - j\frac{\sigma}{\omega\varepsilon_o} \quad (3)$$

where $\varepsilon$ is the dielectric constant at very high frequencies; $\varepsilon_s$ is the dielectric constant at very low frequencies; $\omega$ is the angular frequency (rad $s^{-1}$); $\tau$ is the relaxation time of the dipoles (s); $\sigma$ is the conductivity of the material (Siemens $m^{-1}$); $j$ is the complex operator (i.e. $j = \sqrt{-1}$), and $\varepsilon_o$ is the dielectric permittivity of free space.

Manipulating Eq. (3) to separate it into real and imaginary components yields:

$$\varepsilon = \varepsilon_\infty + \frac{\varepsilon_s - \varepsilon_\infty}{1 + \omega^2\tau^2} - j\left\{\frac{(\varepsilon_s - \varepsilon_\infty)\omega\tau}{1 + \omega^2\tau^2} + \frac{\sigma}{\omega\varepsilon_o}\right\} = \varepsilon' - j\varepsilon'' \quad (4)$$

The relaxation time $\tau$ is a measure of the time required for polar molecules to rotate in response to a changed external electric field, and hence determines the frequency range in which dipole movement occurs. This response time depends on the temperature and physical state of the material.

The dielectric constant ε' affects the wave impedance of the space occupied by the dielectric material [14] and causes reflections at the inter-facial boundary between materials due to changes in the wave impedance of the space occupied by the material. These changes in wave impedance also cause a change in the wavelength of the electromagnetic fields inside the dielectric material, compared with the wavelength in air or vacuum [15]. This change in wavelength affects the propagation velocity of the wave within the material.

The dielectric loss $\varepsilon$" represents the resistive nature of the material [16], which reduces the amplitude of the electromagnetic field and generates heat inside the material.

It is common practice to express the dielectric properties of a material in terms of the relative dielectric constants κ' and κ", which are defined as: $\varepsilon' = \kappa'\varepsilon_o$ and $\varepsilon'' = \kappa''\varepsilon_o$. The general form of the dielectric properties of polar materials resembles the normalised example shown in **Figure 3**.

The dielectric properties of most materials are directly associated with its molecular structure. Debye's basic relationship assumes that the molecules in a material are homogeneous in structure and can be described as "polar". Since few materials can be described in this way, many other equations have been developed to describe frequency-dependent dielectric behaviour.

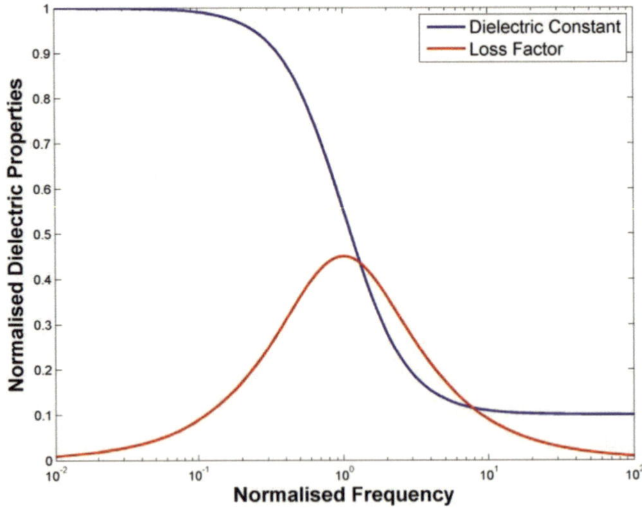


**Figure 3.**
*Normalised dielectric properties of a polar material.*

In many cases, the material may be regarded as a composite or mixture and will exhibit multiple relaxation times. If this is the case then the complex dielectric constant may be represented by a variation of Debye's original equation [17]:

$$\kappa = \kappa_\infty + a\left(\frac{\kappa_s - \kappa_1}{1 + j\omega\tau_1}\right) + b\left(\frac{\kappa_1 - \kappa_2}{1 + j\omega\tau_2}\right) + .. - j\frac{\sigma}{\omega\varepsilon_o} \tag{5}$$

where $\kappa_1$ and $\kappa_2$ are intermediate values of the dielectric constant between the various relaxation periods of $\tau_1$ and $\tau_2$, and $a$ and $b$ are constants related to how much of each component is present in the total material.

## 2.2 Temperature dependence of the dielectric properties

As temperature increases, the electrical dipole relaxation time associated with the material usually decreases, and the loss-factor peak will shift to higher frequencies (**Figure 4**). For many materials, this means that at dispersion frequencies the dielectric constant will increase while the loss factor may either increase or decrease depending on whether the operating frequency is higher or lower than the relaxation frequency [18]. For example, **Table 1** demonstrated how the dielectric properties of some food stuffs, in the microwave band, vary with temperature.

Some food stuffs in **Table 1** follow the predicted trend of increasing dielectric constant as temperature increases; however, it is apparent that the dielectric constant of other entries in **Table 1** decline with increasing temperature rather than increasing with temperature. This is linked to their water content, because the dielectric constant of water at a fixed frequency decreases with increasing temperature (**Figure 4**). **Figure 5** shows how the dielectric properties of water vary with temperature, over a wider range of frequencies.

### *2.2.1 Density dependence of dielectric properties*

Because a dielectric material's influence over electromagnetic waves depends on the amount of the material present in the space occupied by the material, it follows that the density of the material must influence the bulk dielectric properties of

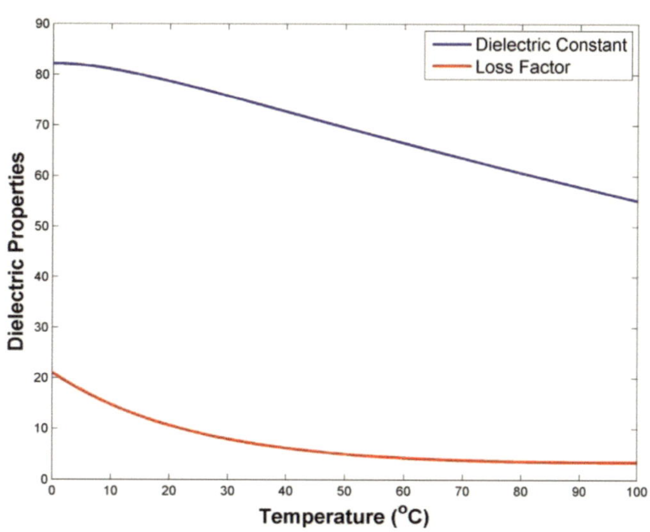


**Figure 4.**
*Dielectric properties of pure water as a function of temperature at 2.45 GHz.*

| Food product | Moisture content | Fat content | Temperature | | | | | |
|---|---|---|---|---|---|---|---|---|
| | | | 50 | | 75 | | 100 | |
| | | | ε' | ε" | ε' | ε" | ε' | ε" |
| Peanut butter | — | 48.5 | 3.1 | 4.1 | 3.2 | 4.5 | 3.5 | 5.0 |
| Ground beef | 50.7 | 31.9 | 39.0 | 10.4 | 32.2 | 11.5 | 31.7 | 12.6 |
| Ham | 69.1 | 4.7 | 66.6 | 47.0 | 87.4 | 57.0 | 101.0 | 60.0 |
| Concentrated orange juice | 57.6 | — | 54.1 | 15.7 | 53.5 | 15.2 | 52.0 | 15.7 |
| Mashed potatoes | 81.3 | 0.9 | 60.6 | 17.4 | 56.3 | 16.5 | 52.9 | 15.8 |
| Peas, cooked | 778.6 | — | 60.8 | 12.6 | 50.5 | 9.7 | 46.6 | 9.1 |

**Table 1.**
*Dielectric properties of some common foods at 2.8 GHz as a function of temperature (source: [19]).*

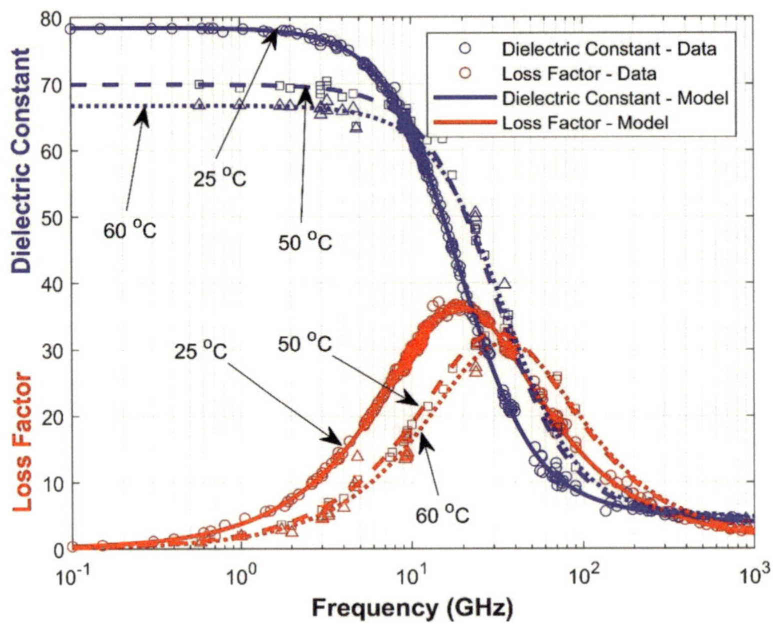


**Figure 5.**
*Dielectric properties of pure water as a function of frequency and temperature (data sources: [20]; model for dielectric properties based on: [21]).*

materials. This is especially true of particulate materials, such as soil, grains or flours [18].

As an example, the dielectric properties of oven dry wood, with the electric field oriented perpendicular to the wood grain, are described by [22]:

$$\kappa' = \kappa_\infty + \frac{(\kappa_o - \kappa_\infty)\left[1 + \omega\tau^{(1-\alpha)}\, Cos\left(\pi\frac{1-\alpha}{2}\right)\right]}{\left[1 + \omega\tau^{2(1-\alpha)} + 2\omega\tau^{(1-\alpha)}\, Cos\left(\pi\frac{1-\alpha}{2}\right)\right]} \tag{6}$$

and

$$\kappa'' = \frac{(\kappa_o - \kappa_\infty)\left[1 + \omega\tau^{(1-\alpha)}\, Cos\left(\pi\frac{1-\alpha}{2}\right)\right]}{2\left[1 + \omega\tau^{2(1-\alpha)} + 2\omega\tau^{(1-\alpha)}\, Cos\left(\pi\frac{1-\alpha}{2}\right)\right]} \tag{7}$$

**Table 2** shows the values of $\kappa_o$ and $\kappa$ as a function of wood density.

### 2.3 Wave propagation

If a material is homogeneous in terms of its electromagnetic properties, it is apparent that an incident electromagnetic wave would be partly reflected at the material boundary and partly transmitted. The transmitted energy would be dissipated due to any losses within the medium. If a plane wave is propagating through space in the x-direction, it can be described by:

$$E(x,t) = E_o e^{j(kx-\omega t)} \tag{8}$$

where, in general, the wavenumber (k) can be described by:

$$k = k_o(\beta + j\alpha) \tag{9}$$

where $k_o$ is the wavenumber of free space:

$$k_o = \frac{2\pi f}{c} \tag{10}$$

Van Remmen, et al. [23] show that the components of the wavenumber are given by:

$$\beta = \sqrt{\kappa' \frac{\sqrt{1 + \left(\frac{\kappa''}{\kappa'}\right)^2} + 1}{2}} \tag{11}$$

and:

$$\alpha = \sqrt{\kappa' \frac{\sqrt{1 + \left(\frac{\kappa''}{\kappa'}\right)^2} - 1}{2}} \tag{12}$$

The term $\alpha$ is associated with wave attenuation with distance travelled through a medium. For free space (or air) $\alpha = 0$ at most frequencies.

For a wave that is perpendicularly incident onto the surface of a material, the reflection coefficient, which is the ratio of the reflected wave amplitude to the incident wave amplitude, is given by:

$$\Gamma = \frac{(\beta_1+\beta_2)(\beta_1-\beta_2)+(\alpha_1+\alpha_2)(\alpha_1-\alpha_2)}{(\beta_1+\beta_2)^2+(\alpha_1+\alpha_2)^2} + j\frac{(\beta_1+\beta_2)(\alpha_1-\alpha_2)-(\beta_1-\beta_2)(\alpha_1+\alpha_2)}{(\beta_1+\beta_2)^2+(\alpha_1+\alpha_2)^2} \tag{13}$$

| Wood density (g cm$^{-3}$) | $\kappa_o$ | $\kappa_\infty$ | Wood density (g cm$^{-3}$) | $\kappa_o$ | $\kappa_\infty$ |
|---|---|---|---|---|---|
| 0.13 | 1.4 | 1.16 | 1.0 | 4.0 | 2.3 |
| 0.2 | 1.6 | 1.2 | 1.2 | 4.8 | 2.5 |
| 0.4 | 2.0 | 1.4 | 1.4 | 6.0 | 2.8 |
| 0.6 | 2.5 | 1.65 | 1.53 | 6.8 | 2.9 |
| 0.8 | 3.2 | 2.03 | | | |

**Table 2.**
*Values of $\kappa_s$ and $\kappa$ for oven dried wood of various densities when the electric field is perpendicular to the wood grain (based on data from: [22]).*

where the subscripts refer to medium 1 and medium 2 across the medium interface. The transmission coefficient, which is the ratio of the transmitted wave amplitude to the incident wave amplitude, is given by:

$$\tau = 2\left\{\frac{\beta_1{}^2 + \beta_1\beta_2 + \alpha_1\alpha_2 + \alpha_1{}^2}{(\beta_1+\beta_2)^2 + (\alpha_1+\alpha_2)^2} + j\frac{\beta_2\alpha_1 - \beta_1\alpha_2}{(\beta_1+\beta_2)^2 + (\alpha_1+\alpha_2)^2}\right\} \tag{14}$$

Therefore, the wave, which propagates across a boundary from one medium (or vacuum) to another, is described by:

$$E(x,t) = E_o\tau \bullet e^{j(k_o\beta x - \omega t)} \bullet e^{-k_o\alpha x} \tag{15}$$

## 3. Transmission through a material

If the electromagnetic wave passes through a material, the wave emerging on the other side will be described by:

$$E(x,t) = E_o\tau_{1,2} \bullet \tau_{2,1} \bullet e^{j(k_o\beta_1 x + k_o\beta_2 L - \omega t)} \bullet e^{-k_o\alpha_2 L} \tag{16}$$

where $\tau_{1,2}$ and $\tau_{2,1}$ are the transmission coefficients of the two material interfaces and $L$ is the thickness of the material through which the wave passes. Therefore, the wave is delayed, phase shifted (because of the complex value of the transmission coefficient) and attenuated by the material, in comparison to a similar wave propagating though free space. This is illustrated in **Figure 6**.

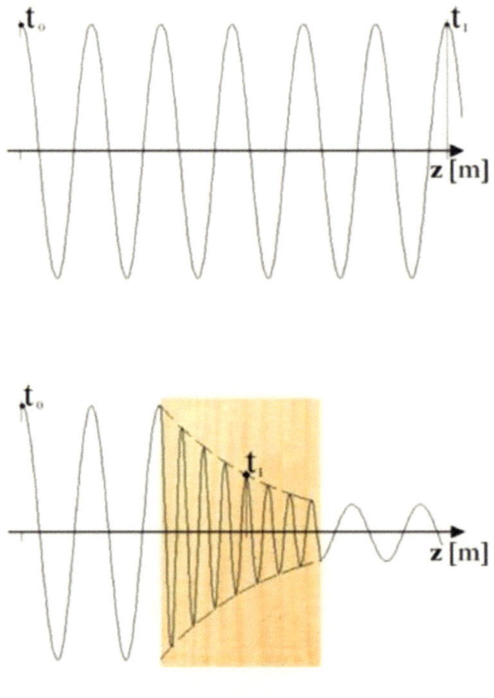


**Figure 6.**
*Schematic of some changes in an electromagnetic wave associated with transmission through a material.*

## 4. Non-invasive detection of internal structures of objects

Wave attenuation, reflections from the material surface, and internal scattering from embedded objects or cavities in the material causes "shadows" on the opposite side of the material from the electromagnetic source. An X-ray image (**Figure 7**) is a good example of how shadows associated with propagation delay and wave attenuation can be used to interpret the internal structures of objects. Effectively the combination of attenuation and phase delay, which are directly linked to the dielectric properties of the material, provide information about the material through which the electromagnetic wave travels.

Wave penetration into any material can be defined by the penetration depth (d), when the ratio of the wave amplitude to initial amplitude is $\frac{1}{e} = 0.3679$:

$$d = \frac{1}{k_o \alpha} \tag{17}$$

The penetration depth of an electromagnetic wave is inversely proportional to the wave frequency and the attenuation factor of the material. If the penetration

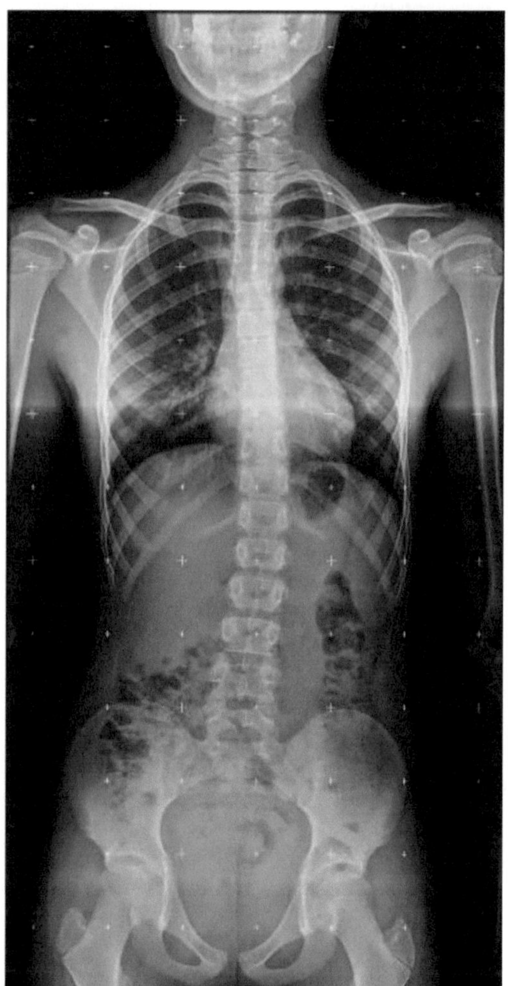

**Figure 7.**
*Example of an X-ray image.*

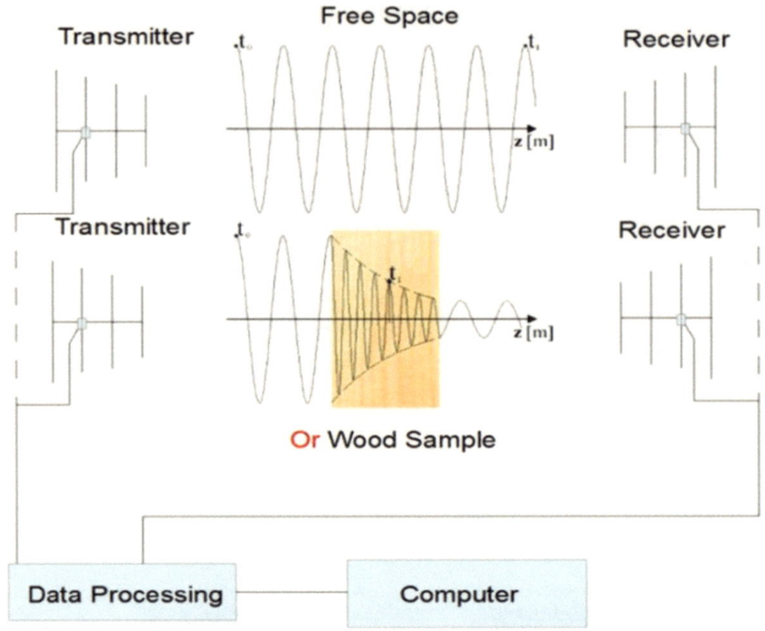


**Figure 8.**
*Schematic of a 'look through' microwave system for assessing the properties of materials.*

depth of a material is far greater than the thickness of the material at a given frequency, the material appears to be 'transparent' to the electromagnetic wave. If the penetration depth of a material is far smaller than the thickness of the material at a given frequency, the material appears to be 'opaque'. If the penetration depth of the material is similar to the thickness of the material at a given frequency, the material may be regarded as 'translucent'.

Although X-rays have been used to investigate the internal features of objects, other frequencies of the electromagnetic spectrum can also be used [24]. For example, microwaves can be used to assess the internal structure of materials, by 'looking through' the objects, as illustrated in **Figure 8**. Such microwave systems have been used to assess moisture content of materials [25–28], detection of decay is timber [29–31], detection of insects in bulk materials such as grains and wood [32–34], assessment of wooden structures of cultural significance [24], and 'see through walls' Wi-Fi imaging [35, 36].

## 5. Energy transfer

Electromagnetic waves can transfer energy from one object to another through open space. Generally, the amount of energy transferred depends on: the intensity of the electromagnetic fields; the frequency of the fields' oscillations; and the dielectric properties of the material. The power dissipated per unit volume in a non-magnetic, uniform materials, exposed to electromagnetic fields can be expressed as [37]:

$$P(x) = 55.63 \times 10^{-12} \bullet f \bullet (\tau \bullet E)^2 \bullet \kappa'' \bullet e^{-2k_o \alpha x} \tag{18}$$

where $f$ is the frequency, $\kappa$" is the dielectric loss factor of the heated material, $k_o$ is the wavenumber of free space, $\alpha$ is the attenuation factor, and $x$ is the distance below the surface of the material. Therefore, more power is dissipated in a material at higher frequencies; however, the wave attenuation factor is higher at higher frequencies and therefore the penetration of the heating into the surface of the material is lower at higher frequencies.

### 5.1 Dielectric heating

Dielectric heating usually takes place in the radio frequency, microwave or millimetre wave bands of the electromagnetic spectrum. Before World War II, there is little evidence of work on dielectric heating; however, Kassner [38] mentions industrial applications of microwave energy in two of his patents on spark-gap microwave generators [38–40]. Unfortunately early studies in radio frequency heating concluded that microwave heating of food stuffs would be most unlikely because the calculated electric field strength required to heat biological materials would approach the breakdown voltage of air [41].

A discovery that microwave energy could heat food by Spencer [42] lead to a series of patents for microwave cooking equipment [43–45]. Radiofrequencies and microwaves interact with all organic materials. The strength of this interaction depends on the dielectric properties of the materials. These dielectric properties are strongly influenced by the amount of water in the material. Absorption of radiofrequency or microwave energy by these dielectric materials generates heat in the material.

The major advantages of dielectric heating are its short start-up, precise control and volumetric heating [46]; however dielectric heating suffers from: uneven temperature distributions [10, 23]; unstable temperatures [47–50]; and rapid moisture movement [51]. The advantage of radio frequency and microwave heating is its volumetric interaction with the heated material as the electromagnetic energy is absorbed by the material and manifested as heat [52]. This means that the heating behaviour is not restricted by the thermal diffusivity of the heated material.

In industry, dielectric heating is used for drying [46, 53–55], oil extraction from tar sands, cross-linking of polymers, metal casting [46], medical applications [56], pest control [32], enhancing seed germination [57], and solvent free chemistry [58].

The temperature response of the material, other than on the surface, is limited by the coefficient of simultaneous heat and moisture movement [51]. If for any reason the local diffusion rate is much less than the electromagnetic power dissipation rate, the local temperature will increase rapidly. With increasing temperature, the properties of the material change. If such changes lead to the acceleration of electromagnetic power dissipation at this local point, the temperature will increase more rapidly. The result of such a positive feedback is the formation of a hot spot, which is a local thermal runaway [59].

Thermal runaway, which manifests itself as a sudden temperature rise due to small increases in the applied electromagnetic power, is very widely documented [48, 60]. It has also been reported after some time of steady heating at fixed power levels and is usually attributed to temperature dependent dielectric and thermal properties of the material [48, 60].

### 5.2 Hot body radiation energy transfer

Any object that is above zero degrees Kelvin will radiate energy in the form of electromagnetic photons. The German physicist, Max Planck (1858–1947), deduced that the radiation spectral density ($\rho$) given off from a hot object depended on the wavelength of interest and the temperature of the object. This spectral density can be described by:

$$\rho = \frac{2hc^2}{\lambda^5\left\{e^{\frac{hc}{\lambda kT}} - 1\right\}} \quad (19)$$

where $h$ is Planck's constant ($6.6256 \times 10^{-34}$ J s), $c$ is the speed of light, $\lambda$ is the electromagnetic wavelength of interest, $k$ is Boltzmann's constant ($1.38054 \times 10^{-23}$ J K$^{-1}$), and $T$ is the temperature in Kelvin. A typical set of spectral distributions for different temperatures is shown in **Figure 9**.

The wavelength at which peak radiation intensity occurs can be found by differentiating Planck's equation and setting the derivative equal to zero. Therefore, the wavelength of peak radiation is determined by:

$$\lambda_p \approx \frac{hc}{5kT} \quad (20)$$

where $\lambda_p$ is the peak radiation wavelength (m). At room temperature, or above, the wavelength of peak radiation will be in the micrometre range (~10 μm), which is in the long-wavelength infrared band (**Table 3**). The penetration of electromagnetic energy into materials is limited by the wavelength and the dielectric properties of the material [61], as pointed out in Eq. (17). The penetration depth of any radiation from objects at room temperature, or above, will be in the nanometre

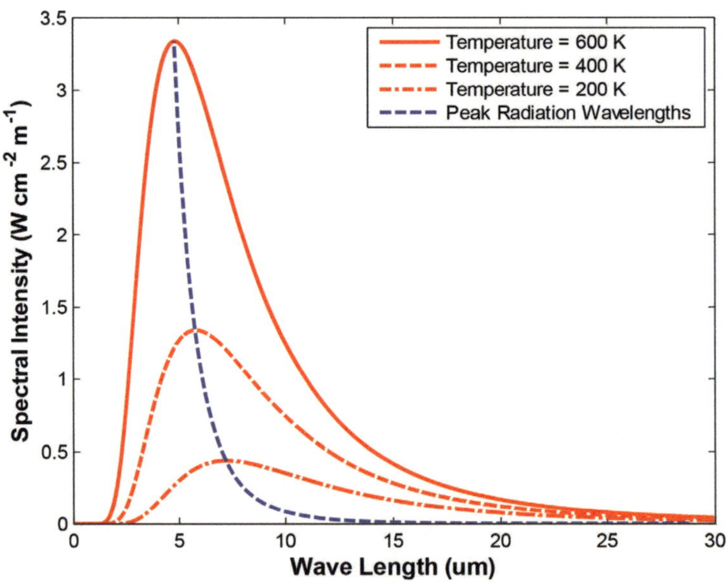


**Figure 9.**
*Radiative spectral density at different temperatures as a function of temperature and wavelength.*

| Division name | Abbreviation | Wavelength (μm) | Temperature (K) |
|---|---|---|---|
| Near infrared | NIR | 0.75–1.4 | 3964–2070 |
| Short-wavelength Infrared | SWIR | 1.4–3.0 | 2070–966 |
| Mid-wavelength Infrared | MWIR | 3.0–8.0 | 966–362 |
| Long-wavelength Infrared | LWIR | 8.0–15.0 | 362–193 |
| Far infrared | FIR | 15.0–1000 | 193–3 |

**Table 3.**
*A commonly used sub-division scheme for the infrared part of the electromagnetic spectrum.*

range; therefore, this form of radiative energy transfer must be regarded as a surface phenomenon, where further energy transfer from the surface into the material occurs via internal conduction and convection.

The total radiated power can be determined by integrating Planck's equation across all wavelengths for a temperature to yield the Stefan-Boltzmann equation. The power transferred from an object at one temperature to another object at a lower temperature is given by [62]:

$$q = \varepsilon\sigma A\left(T_A^4 - T_{\mathrm{P}}^4\right) \tag{21}$$

where $q$ is the radiation power transferred (W); $\varepsilon$ is the surface emissivity of the radiator material; $\sigma$ is the Stefan-Boltzmann constant ($5.6704 \times 10^{-8}\,\mathrm{J\,s^{-1}\,m^{-2}\,K^{-4}}$); $A$ is the surface area of the heated object ($\mathrm{m}^2$); $T_A$ is the temperature of the infrared source (K); and $T_p$ is the temperature of the material being heated (K). In the case of a normal object the power transfer is reduced by a factor $\varepsilon$, which depends on the properties of the object's surface. This factor is referred to as the emissivity of the surface.

### 5.3 Thermal imaging

Brightness temperature is the temperature that a black body, in thermal equilibrium with its surroundings, would need to have in order to duplicate the observed electromagnetic wave intensity, at a known wavelength. The brightness temperature of a body can be determined by rearranging Planck's equation to find $T_b$ for a given spectral density value, at a wavelength $\lambda$:

$$T_b = \frac{\mathrm{hc}^2}{\lambda k \bullet \ln\left(\frac{2\mathrm{hc}}{\rho\lambda^5} + 1\right)} \tag{22}$$

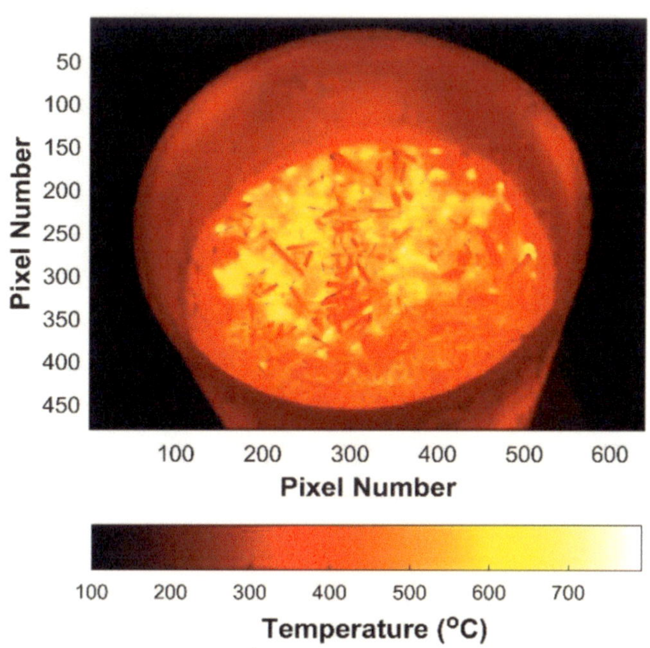


**Figure 10.**
*Thermal image of a microwave heated sample of biosolids.*

The real surface temperature of an object can be determined by dividing the brightness temperature by the surface emissivity of the object being assessed. Since the emissivity is a value between 0 and 1, the real temperature will be greater than or equal to the brightness temperature. This is effectively how remote thermal sensors and thermal imaging systems operate (**Figure 10**).

## 6. Conclusion

Space-time behaves like a dielectric, allowing electromagnetic waves to propagate through it. The inclusion of material objects in space alters the fundamental dielectric properties of space-time in such a way that electromagnetic energy is reflected from the surface of material objects, transmitted through material objects and absorbed by material objects. These interactions, along with other changes in the electromagnetic wave propagation, allow: information about the space occupied by the material object to be non-invasively acquired; or facilitates the transfer of energy from the electromagnetic wave to the material object. In a practical sense, these interactions between electromagnetic waves and material objects can facilitate remote sensing and energy transfer over very long distances.

## Author details

Graham Brodie
Faculty of Veterinary and Agricultural Sciences, The University of Melbourne, Dookie Campus, Dookie, Australia

*Address all correspondence to: grahamb@unimelb.edu.au

## References

[1] Dirac PAM. The quantum theory of the emission and absorption of radiation. Proceedings of the Royal Society of London. Series A, Containing Papers of a Mathematical and Physical Character. 1927;**114**:243-265

[2] Einstein A. The advent of the quantum theory. Science. 1951;**113**:82-84

[3] Einstein A. Relativity: The Special and General Theory. London: Methuen & Co Ltd; 1916

[4] Chappell JM, Iqbal A, Abbott D. A Simplified Approach to Electromagnetism Using Geometric Algebra. Ithaca, NY, USA: Cornell University; 2010

[5] Evans MW. The spinning and curving of spacetime: The electromagnetic and gravitational fields in the Evans field theory. Foundations of Physics Letters. 2005;**18**:431-454

[6] International Telecommunication Union. Spectrum Management for a Converging World: Case Study on Australia. Geneva, Switzerland: International Telecommunication Union; 2004

[7] Commonwealth Department of Transport and Communications. Australian Radio Frequency Spectrum Allocations. Canberra: Commonwealth Department of Transport and Communications; 1991

[8] Adamski W, Kitlinski M. On measurements applied in scientific researches of microwave heating processes. Measurement Science Review. 2001;**1**:199-203

[9] Lundgren N. Modelling Microwave Measurements. Wood Department of Skellefteå Campus, Division of Wood Science and Technology, Luleå University of Technology; 2005

[10] Brodie G. The influence of load geometry on temperature distribution during microwave heating. Transactions of the American Society of Agricultural and Biological Engineers. 2008;**51**: 1401-1413

[11] Brodie GI. Innovative Wood Drying: Applying Microwave and Solar Technologies to Wood Drying. Saarbruecken, Germany: VDM Verlag; 2008. 120 p

[12] Amador F, Carpenter P, Evans GJ, Evans MW, Felker L, Guala-Valverde J, et al. Dielectric theory of ECE spacetime. In: Evans MW, editor. Generally Covariant Unified Field Theory—The Geometrization of Physics-Volume III. London: Abramis; 2006. pp. 23-31

[13] Debye P. Polar Molecules. New York: Chemical Catalog; 1929

[14] Singh RP, Heldman DR. Introduction to Food Engineering. New York: Academic Press; 1993

[15] Montoro T, Manrique E, Gonzalez-Reviriego A. Measurement of the refracting index of wood for microwave radiation. Holz als Roh- und Werkstoff. 1999;**57**:295-299

[16] Smith RJ. Circuits, Devices and Systems. New York: Wiley International; 1976

[17] Kuang W, Nelson SO. Dielectric relaxation characteristics of fresh fruits and vegetables from 3 to 20 GHz. Journal of Microwave Power and Electromagnetic Energy. 1997;**32**: 114-122

[18] Nelson SO. Dielectric properties of agricultural products: Measurements and applications. IEEE Transactions on Electrical Insulation. 1991;**26**:845-869

[19] Ohlsson T, Bengtsson NE. Dielectric food data for microwave sterilization processing. The Journal of Microwave Power. 1975;**10**:93-108

[20] Ellison WJ, Lamkaouchi K, Moreau JM. Water: A dielectric reference. Journal of Molecular Liquids. 1996;**68**: 171-279

[21] Meissner T, Wentz FJ. The complex dielectric constant of pure and sea water from microwave satellite observations. IEEE Transactions on Geoscience and Remote Sensing. 2004;**42**: 1836-1849

[22] Torgovnikov GI. Dielectric Properties of Wood and Wood-Based Materials. Berlin: Springer-Verlag; 1993

[23] Van Remmen HHJ, Ponne CT, Nijhuis HH, Bartels PV, Herkhof PJAM. Microwave heating distribution in slabs, spheres and cylinders with relation to food processing. Journal of Food Science. 1996;**61**:1105-1113

[24] Brodie G, Harris E, Farrell P, Tse NA, Roberts A, Kvansakul J. In-situ, noninvasive investigation of an outdoor wooden sculpture. In: Proceedings of ICOM-CC 17th Triennial Conference. 2014. pp. 1-9

[25] Kim K, Kim J, Lee SS, Noh SH. Measurement of grain moisture content using microwave attenuation at 10.5 GHz and moisture density. IEEE Transactions on Instrumentation and Measurement. 2002;**51**:72-77

[26] Kim KB, Kim JH, Lee CJ, Noh SH, Kim MS. Simple instrument for moisture measurement in grain by free-space microwave transmission. Transactions of the American Society of Agricultural and Biological Engineers. 2006;**49**:1089-1093

[27] Moschler WW, Hanson GR, Gee TF, Killough SM, Wilgen JB. Microwave moisture measurement system for lumber drying. Forest Products Journal. 2007;**57**:69-74

[28] Rouleau JF, Goyette J, Bose TK, Frechette MF. Performance of a microwave sensor for the precise measurement of water vapor in gases. IEEE Transactions on Dielectrics and Electrical Insulation. 2000;**7**:825-831

[29] Brodie G, Ahmed B. Microwave technologies for detecting termites and decay in timber. In: Proceedings of The 13th International Conference on Microwave and High Frequency Heating. 2011. pp. 131-134

[30] Brodie G, Ahmed B, Jacob MV. Using microwave transmission to detect decay in wood. In: Proceedings of 19th Annual International Conference on Composites and Nano Engineering. 2011

[31] Brodie G, Ahmed BM, Jacob MV. Detection of decay in wood using microwave characterization. In: Proceedings of 2011 Asia-Pacific Microwave Conference Proceedings (APMC). 2011. pp. 1754-1757

[32] Nelson SO. Insect-control possibilities of electromagnetic energy. Cereal Science Today. 1972;**17**:377-387

[33] Nelson SO. Potential agricultural applications of RF and microwave energy. Transactions of the ASAE. 1987; **30**:818-831

[34] Nelson SO. Review and assessment of radio-frequency and microwave energy for stored-grain insect control. Transactions of the ASAE. 1996;**39**: 1475-1484

[35] Yunqiang Y, Fathy AE. See-through-wall imaging using ultra wideband short-pulse radar system. In: Proceedings of 2005 IEEE Antennas and Propagation Society International Symposium. Vol. 3B. 2005. pp. 334-337

[36] Adib F, Katabi D. See through walls with WiFi! In: Proceedings of the ACM SIGCOMM 2013 Conference on SIGCOMM. 2013. pp. 75-86

[37] Brodie G. Applications of microwave heating in agricultural and forestry related industries. In: Cao W, editor. The Development and Application of Microwave Heating. Rijeka, Croatia: InTech; 2012. pp. 45-78

[38] Kassner EEW. Process for altering the energy content of dipolar substances. United States Patent 2089966; 1937

[39] Kassner EEW. Apparatus for the generation of short electromagnetic waves. United States Patent 2094602; 1937

[40] Kassner EEW. Apparatus for generating and applying ultrashort electromagnetic waves. United States Patent 2109843; 1938

[41] Shaw TM, Galvin JA. High-frequency-heating characteristics of vegetable tissues determined from electrical-conductivity measurements. Proceedings of the IRE. 1949;**37**:83-86

[42] Murray D. Percy Spencer and his itch to know. Reader's Digest. 1958

[43] Spencer PL. Magnetron anode structure. United States Patent 2417789; 1947

[44] Spencer PL. Prepared food article and method of preparing. United States Patent 2480679; 1949

[45] Spencer PL. Electronic cooking. United States Patent 2582174; 1952

[46] Ayappa KG, Davis HT, Crapiste G, Davis EJ, Gordon J. Microwave heating: An evaluation of power formulations. Chemical Engineering Science. 1991;**46**: 1005-1016

[47] Vriezinga CA. Thermal runaway and bistability in microwave heated isothermal slabs. Journal of Applied Physics. 1996;**79**:1779-1783

[48] Vriezinga CA. Thermal runaway in microwave heated isothermal slabs, cylinders, and spheres. Journal of Applied Physics. 1998;**83**:438-442

[49] Vriezinga CA. Thermal profiles and thermal runaway in microwave heated slabs. Journal of Applied Physics. 1999; **85**:3774-3779

[50] Vriezinga CA, Sanchez-Pedreno S, Grasman J. Thermal runaway in microwave heating: A mathematical analysis. Applied Mathematical Modelling. 2002;**26**:1029-1038

[51] Brodie G. Simultaneous heat and moisture diffusion during microwave heating of moist wood. Applied Engineering in Agriculture. 2007;**23**: 179-187

[52] Metaxas AC, Meredith RJ. Industrial Microwave Heating. London: Peter Peregrinus; 1983

[53] Antti AL, Perre P. A microwave applicator for on line wood drying: Temperature and moisture distribution in wood. Wood Science and Technology. 1999;**33**:123-138

[54] Hasna A, Taube A, Siores E. Moisture monitoring of corrugated board during microwave processing. Journal of Electromagnetic Waves and Applications. 2000;**14**:1563

[55] Zielonka P, Dolowy K. Microwave drying of spruce: Moisture content, temperature and heat energy distribution. Forest Products Journal. 1998;**48**:77-80

[56] Bond EJ, Li X, Hagness SC, Van Veen BD. Microwave imaging via space-time beamforming for early detection of breast cancer. IEEE Transaction on

Antennas and Propagation. 2003;**51**: 1690-1705

[57] Nelson SO, Stetson LE. Germination responses of selected plant species to RF electrical seed treatment. Transactions of the ASAE. 1985;**28**:2051-2058

[58] Arrieta A, Otaegui D, Zubia A, Cossio FP, Diaz-Ortiz A, delaHoz A, et al. Solvent-free thermal and microwave-assisted [3+2] cycloadditions between stabilized azomethine ylides and nitrostyrenes. An experimental and theoretical study. Journal of Organic Chemistry. 2007;**72**: 4313-4322

[59] Wu X. Experimental and Theoretical Study of Microwave Heating of Thermal Runaway Materials Mechanical Engineering. Virginia Polytechnic Institute and State University; 2002

[60] Zielonka P, Gierlik E. Temperature distribution during conventional and microwave wood heating. Holz als Roh- und Werkstoff. 1999;**57**:247-249

[61] Vollmer M. Physics of the microwave oven. Physics Education. 2004;**39**:74-81

[62] Holman JP. Heat Transfer. New York: McGraw-Hill; 1997

Chapter 5

# Two Systems of Maxwell's Equations and Two Corresponding Systems of Wave Equations in a Rotating Dielectric Medium

*Evgeny A. Bondarenko*

## Abstract

In this chapter, on the base of two basic systems of Maxwell's equations for electromagnetic field vectors $\vec{E}$ and $\vec{B}$ in a uniformly rotating dielectric medium, the two corresponding systems of wave equations have been derived (to the first order in an angular velocity $\Omega$). From their comparative analysis, it can be seen that the structure of the wave equations for electromagnetic field vectors in the first system is asymmetrical with respect to $\Omega$, while the structure of such equations in the second one is symmetrical. On this basis, it can be concluded that if the principle of symmetry is accepted as a criterion for selection, then second system of wave equations (and, therefore, corresponding the second set of Maxwell's equations) for vectors $\vec{E}$ and $\vec{B}$ may be preferred.

**Keywords:** Maxwell's equations, wave equations, Sagnac effect, fiber-optic gyro, ring laser gyro

## 1. Introduction

In order to develop the theory of fiber-optic gyro (or, e.g., ring laser gyro with resonator containing a dielectric medium with index of refraction $n$), one needs to have a system of Maxwell's equations and corresponding system of wave equations for electromagnetic field vectors $\vec{E}$ and $\vec{B}$ which are written in a frame of reference which uniformly rotates in an inertial frame with angular velocity $\Omega$ ($\Omega = |\vec{\Omega}|$). Since the module $v$ of vector $\vec{v} = \vec{\Omega} \times \vec{r}$ of a tangential velocity of such rotating device is much smaller than the speed of light, it is sufficient for these systems to be linear in $v$ or, equivalently, in $\Omega$.

From the literature, it can be seen that there are mainly two basic systems of Maxwell's equations for electromagnetic field vectors $\vec{E}$ and $\vec{B}$ derived from the first principles and written for the case of a uniformly rotating dielectric medium. These two systems of equations are in good agreement with the experiments conducted for ring optical interferometers, fiber-optic gyros, and ring laser gyros containing in their arms the gas discharge tubes with Brewster's windows. Both systems are based on the Galilean description of rotation: $x = x^0 \cos\Omega t^0 + y^0 \sin\Omega t^0, y = -x^0 \sin\Omega t^0 + y^0 \cos\Omega t^0, z = z^0, t = t^0$ (superscript '0' refers to the inertial frame).

In the absence of free charges and currents, the first system, which was first obtained in work [1] from the formalism of the theory of general relativity, may be written in the form (we keep the terms only up to first order in $\Omega$):

$$\vec{\nabla} \times \vec{E} + \frac{\partial \vec{B}}{\partial t} = 0, \tag{1}$$

$$\vec{\nabla} \cdot \vec{B} = 0, \tag{2}$$

$$\vec{\nabla} \times \left(\vec{B} - \frac{1}{c^2}\,\vec{v} \times \vec{E}\right) - \frac{1}{c^2}\frac{\partial}{\partial t}\left(n^2\,\vec{E} - \vec{v} \times \vec{B}\right) = 0, \tag{3}$$

$$\vec{\nabla} \cdot \left(\vec{E} - \frac{1}{n^2}\,\vec{v} \times \vec{B}\right) = 0, \tag{4}$$

and the second one, which was first obtained in work [2] on the base of the use of the tetrad method in this theory, may be presented as

$$\vec{\nabla} \times \vec{E} + \frac{\partial}{\partial t}\left(\vec{B} + \frac{1}{c^2}\,\vec{v} \times \vec{E}\right) = 0, \tag{5}$$

$$\vec{\nabla} \cdot \left(\vec{B} + \frac{1}{c^2}\,\vec{v} \times \vec{E}\right) = 0, \tag{6}$$

$$\vec{\nabla} \times \vec{B} - \frac{1}{c^2}\frac{\partial}{\partial t}\left(n^2\,\vec{E} - \vec{v} \times \vec{B}\right) = 0, \tag{7}$$

$$\vec{\nabla} \cdot \left(\vec{E} - \frac{1}{n^2}\,\vec{v} \times \vec{B}\right) = 0. \tag{8}$$

Systems of Maxwell's Eqs. (1–4) and (5–8) are written here in units of the SI. They contain only two electromagnetic field vectors $\vec{E}$ and $\vec{B}$. In the named original works [1] (see expressions (5), (14b) therein) and [2] (see relations (6.11), (6.12) therein), these systems contain, besides vectors $\vec{E}$ and $\vec{B}$, two more electromagnetic field vectors $\vec{D}$ and $\vec{H}$ (which must be excluded). Systems (1–4) and (5–8) can be also found, for example, in subsequent works [3–8] and [9], respectively. In the case of vacuum (when $n = 1$), sets (1–4) and (5–8) are transformed into the ones known correspondingly from works [10] and [11].

In the above systems, all the quantities are specified by the formulas

$$\begin{gathered}\vec{\nabla} = \hat{x}(\partial/\partial x) + \hat{y}(\partial/\partial y) + \hat{z}(\partial/\partial z),\ \vec{E} = E_x\,\hat{x} + E_y\,\hat{y} + E_z\,\hat{z},\ \vec{B} = B_x\,\hat{x} + B_y\,\hat{y} + B_z\,\hat{z},\\ \vec{\Omega} = \Omega_x\,\hat{x} + \Omega_y\,\hat{y} + \Omega_z\,\hat{z},\ \vec{r} = x\hat{x} + y\hat{y} + z\hat{z},\ \vec{v} = \vec{\Omega} \times \vec{r} = v_x\,\hat{x} + v_y\,\hat{y} + v_z\,\hat{z},\\ v_x = \Omega_y z - \Omega_z y, v_y = \Omega_z x - \Omega_x z, v_z = \Omega_x y - \Omega_y x\end{gathered} \tag{9}$$

where $\hat{x}, \hat{y}$, and $\hat{z}$ are the unit vectors which form an orthogonal coordinate basis $\{\hat{x}\hat{y}\hat{z}\}$ of a rotating frame; $E_x$, $E_y$, and $E_z$ and $B_x$, $B_y$, and $B_z$ are the components of vectors $\vec{E}$ and $\vec{B}$ in this basis; $\vec{\Omega}$ is the vector of angular velocity with which the basis $\{\hat{x}\hat{y}\hat{z}\}$ rotates in the inertial frame; $\Omega_x$, $\Omega_y$, and $\Omega_z$ are the components of vector $\vec{\Omega}$; $\vec{r}$ is the radius vector of the given observation point in basis $\{\hat{x}\hat{y}\hat{z}\}$; $x, y$, and $z$

are the components of vector $\vec{r}$; $\vec{v}$ is the vector of linear tangential velocity of the observation point calculated in the inertial frame; $v_x$,$v_y$, and $v_z$ are the components of vector $\vec{v}$; $n = (\varepsilon_r \mu_r)^{1/2}$ is the index of refraction of a dielectric medium; and $\varepsilon_r$ and $\mu_r$ are the relative permittivity and permeability of a medium, respectively.

Remark about systems (1–4) and (5–8)

⊲ 1. According to the textbook [12], in a uniformly rotating frame of reference with spatial rectangular coordinates $x, y, z$ and time coordinate $t$, the quadratic form $ds^2$ may be presented as $ds^2 = g_{ik}\, dx^i\, dx^k$ where $x^0 = ct$, $x^1 = x$, $x^2 = y$, and $x^3 = z$. In general, when $\vec{\Omega} = \Omega_x \hat{x} + \Omega_y \hat{y} + \Omega_z \hat{z}$, the nonzero components of a space–time metric tensor (with determinant $g = -1$) of such rotating frame may be written as $g_{00} = 1 - v^2/c^2$, $g_{11} = g_{22} = g_{33} = -1$, $g_{01} = g_{10} = -v_x/c$, $g_{02} = g_{20} = -v_y/c$, and $g_{03} = g_{30} = -v_z/c$. Then, components $\kappa_{\alpha\beta} = -g_{\alpha\beta} + g_{0\alpha}\, g_{0\beta}/g_{00}$ of a spatial metric tensor [with determinant $\kappa = \left(1 - v^2/c^2\right)^{-1}$] of such frame will be $\kappa_{11} = 1 + \kappa v_x^2/c^2$, $\kappa_{22} = 1 + \kappa v_y^2/c^2$, $\kappa_{33} = 1 + \kappa v_z^2/c^2$, $\kappa_{12} = \kappa_{21} = \kappa v_x v_y/c^2$, $\kappa_{13} = \kappa_{31} = \kappa v_x v_z/c^2$, and $\kappa_{23} = \kappa_{32} = \kappa v_y v_z/c^2$. From these formulas for $\kappa_{\alpha\beta}$ and $\kappa$, it follows that the spatial metric tensor of a rotating frame of reference, in a linear with respect to $\Omega$ approximation, has a diagonal form with nonzero elements $\kappa_{11} = \kappa_{22} = \kappa_{33} = 1$ and its determinant $\kappa = 1$. Therefore, in such approximation, geometry of space in a rotating frame of reference remains Euclidean (flat space). So, spatial rectangular coordinates $x$, $y$, and $z$ (or, e.g., cylindrical coordinates $\rho$, $\phi$, $z$) of the given observation point in this frame have their usual sense, and the operator $\vec{\nabla}$ in systems (1–4) and (5–8) may be used in usual way. ⊳.

As one can see, the above two systems of Maxwell's Eqs. (1–4) and (5–8) for electromagnetic field vectors $\vec{E}$ and $\vec{B}$ are not identical: system (1–4) has asymmetrical structure with respect to $\Omega$ in a sense that rotation manifests itself only in the third and fourth equations but not in the first and second ones; system (5–8) has symmetrical structure with respect to $\Omega$ because rotation manifests itself in all four equations. The reason of such difference between these two systems is that they were obtained in works [1, 2] with the help of two qualitatively different theoretical approaches.

In this situation, one may ask the following questions: (1) what will the form of the two corresponding systems of wave equations for the named vectors in first and second cases be? (2) Which system of such wave equations (first or second) is preferred? The answers to these questions are not given in the literature.

So, the task of this research is to derive the wave equations for vectors $\vec{E}$ and $\vec{B}$— at first on the base of system of Maxwell's Eqs. (1–4) and then on the base of system (5–8). The results obtained in both cases must be compared. All calculations must be performed with accuracy approximated to the first order in $v$ ($v = |\ \vec{v}\ |$) or, equivalently, $\Omega$ ($\Omega = |\ \vec{\Omega}\ |$).

## 2. Auxiliary relations

In this section, we are going to present some useful formulas for the quantities $\vec{\nabla} \times \left(\vec{\nabla} \times \vec{G}\right)$, $\vec{\nabla} \times \left(\vec{v} \times \vec{G}\right)$, $\vec{\nabla} \cdot \left(\vec{v} \times \vec{G}\right)$, $\vec{\nabla}\left(\vec{v} \cdot \vec{G}\right)$, and $\nabla^2 \left(\vec{v} \times \vec{G}\right)$ which involve vectors $\vec{v} = \vec{\Omega} \times \vec{r}$ and $\vec{G}$ ($\vec{G} = \vec{E}$, $\vec{B}$).

A. Consider the term $\vec{\nabla} \times \left(\vec{\nabla} \times \vec{G}\right)$. It is known (see, e.g., handbook [13]) that

$$\vec{\nabla}\times\left(\vec{\nabla}\times\vec{G}\right)=-\vec{\nabla}^2\vec{G}+\vec{\nabla}\left(\vec{\nabla}\cdot\vec{G}\right). \tag{10}$$

B. Consider the identity $\vec{\nabla}\times\left(\vec{v}\times\vec{G}\right)=\left(\vec{G}\cdot\vec{\nabla}\right)\vec{v}-\left(\vec{v}\cdot\vec{\nabla}\right)\vec{G}+\vec{v}\left(\vec{\nabla}\cdot\vec{G}\right)-\vec{G}\left(\vec{\nabla}\cdot\vec{v}\right)$. In the case $\vec{v}=\vec{\Omega}\times\vec{r}$, we have $\left(\vec{G}\cdot\vec{\nabla}\right)\vec{v}=\vec{\Omega}\times\vec{G}$, $\vec{\nabla}\cdot\vec{v}=0$, and, to the first order in $\Omega$, $\vec{v}\left(\vec{\nabla}\cdot\vec{G}\right)=0$. Therefore

$$\vec{\nabla}\times\left(\vec{v}\times\vec{G}\right)=-\left(\vec{v}\cdot\vec{\nabla}\right)\vec{G}+\vec{\Omega}\times\vec{G}. \tag{11}$$

C. Consider the identity $\vec{\nabla}\cdot\left(\vec{v}\times\vec{G}\right)=\vec{G}\cdot\left(\vec{\nabla}\times\vec{v}\right)-\vec{v}\cdot\left(\vec{\nabla}\times\vec{G}\right)$. Since $\vec{\nabla}\times\vec{v}=2\,\vec{\Omega}$,

$$\vec{\nabla}\cdot\left(\vec{v}\times\vec{G}\right)=2\,\vec{\Omega}\cdot\vec{G}-\vec{v}\cdot\left(\vec{\nabla}\times\vec{G}\right). \tag{12}$$

D. Consider the identity $\vec{\nabla}\left(\vec{v}\cdot\vec{G}\right)=\left(\vec{v}\cdot\vec{\nabla}\right)\vec{G}+\left(\vec{G}\cdot\vec{\nabla}\right)\vec{v}+\vec{v}\times\left(\vec{\nabla}\times\vec{G}\right)+\vec{G}\times\left(\vec{\nabla}\times\vec{v}\right)$. Since $\left(\vec{G}\cdot\vec{\nabla}\right)\vec{v}=\vec{\Omega}\times\vec{G}$, $\vec{\nabla}\times\vec{v}=2\,\vec{\Omega}$, and $\vec{G}\times\left(\vec{\nabla}\times\vec{v}\right)=-2\left(\vec{\Omega}\times\vec{G}\right)$,

$$\vec{\nabla}\left(\vec{v}\cdot\vec{G}\right)=\left(\vec{v}\cdot\vec{\nabla}\right)\vec{G}-\vec{\Omega}\times\vec{G}+\vec{v}\times\left(\vec{\nabla}\times\vec{G}\right). \tag{13}$$

E. Consider the vector $\vec{\nabla}^2\left(\vec{v}\times\vec{G}\right)$. There is the formula

$$\vec{\nabla}^2\left(\vec{v}\times\vec{G}\right)=\vec{\nabla}^2\left(\vec{v}\times\vec{G}\right)_x\hat{x}+\vec{\nabla}^2\left(\vec{v}\times\vec{G}\right)_y\hat{y}+\vec{\nabla}^2\left(\vec{v}\times\vec{G}\right)_z\hat{z}. \tag{14}$$

Let us first calculate the projection $\vec{\nabla}^2\left(\vec{v}\times\vec{G}\right)_x$ of this vector onto the axis $\hat{x}$. Taking into account (9), we have

$$\vec{\nabla}^2\left(\vec{v}\times\vec{G}\right)_x=\left(\frac{\partial^2}{\partial x^2}+\frac{\partial^2}{\partial y^2}+\frac{\partial^2}{\partial z^2}\right)(v_y G_z-v_z G_y), \tag{15}$$

or

$$\vec{\nabla}^2\left(\vec{v}\times\vec{G}\right)_x=\left(\frac{\partial^2}{\partial x^2}+\frac{\partial^2}{\partial y^2}+\frac{\partial^2}{\partial z^2}\right)\left[(\Omega_z x-\Omega_x z)\,G_z-(\Omega_x y-\Omega_y x)\,G_y\right]. \tag{16}$$

After the calculation, we get

$$\begin{aligned}\vec{\nabla}^2\left(\vec{v}\times\vec{G}\right)_x&=v_y\left(\frac{\partial^2 G_z}{\partial x^2}+\frac{\partial^2 G_z}{\partial y^2}+\frac{\partial^2 G_z}{\partial z^2}\right)-v_z\left(\frac{\partial^2 G_y}{\partial x^2}+\frac{\partial^2 G_y}{\partial y^2}+\frac{\partial^2 G_y}{\partial z^2}\right)\\&-2\Omega_x\left(\frac{\partial G_y}{\partial y}+\frac{\partial G_z}{\partial z}\right)+2\left(\Omega_y\frac{\partial G_y}{\partial x}+\Omega_z\frac{\partial G_z}{\partial x}\right).\end{aligned} \tag{17}$$

Let us add to the right-hand side of (17) the following terms: $-2\Omega_x\,(\partial G_x/\partial x)$ and $+2\Omega_x\,(\partial G_x/\partial x)$. Then

$$\vec{\nabla}^2\left(\vec{v}\times\vec{G}\right)_x = v_y\left(\frac{\partial^2 G_z}{\partial x^2}+\frac{\partial^2 G_z}{\partial y^2}+\frac{\partial^2 G_z}{\partial z^2}\right) - v_z\left(\frac{\partial^2 G_y}{\partial x^2}+\frac{\partial^2 G_y}{\partial y^2}+\frac{\partial^2 G_y}{\partial z^2}\right)$$
$$-2\Omega_x\left(\frac{\partial G_x}{\partial x}+\frac{\partial G_y}{\partial y}+\frac{\partial G_z}{\partial z}\right)+2\left(\Omega_x\frac{\partial G_x}{\partial x}+\Omega_y\frac{\partial G_y}{\partial x}+\Omega_z\frac{\partial G_z}{\partial x}\right), \tag{18}$$

or

$$\vec{\nabla}^2\left(\vec{v}\times\vec{G}\right)_x = \left[\vec{v}\times\left(\vec{\nabla}^2\vec{G}\right)\right]_x - 2\Omega_x\left(\vec{\nabla}\cdot\vec{G}\right)+2\frac{\partial}{\partial x}\left(\vec{\Omega}\cdot\vec{G}\right). \tag{19}$$

Similarly, we may obtain

$$\vec{\nabla}^2\left(\vec{v}\times\vec{G}\right)_y = \left[\vec{v}\times\left(\vec{\nabla}^2\vec{G}\right)\right]_y - 2\Omega_y\left(\vec{\nabla}\cdot\vec{G}\right)+2\frac{\partial}{\partial y}\left(\vec{\Omega}\cdot\vec{G}\right), \tag{20}$$

$$\vec{\nabla}^2\left(\vec{v}\times\vec{G}\right)_z = \left[\vec{v}\times\left(\vec{\nabla}^2\vec{G}\right)\right]_z - 2\Omega_z\left(\vec{\nabla}\cdot\vec{G}\right)+2\frac{\partial}{\partial z}\left(\vec{\Omega}\cdot\vec{G}\right). \tag{21}$$

Therefore

$$\vec{\nabla}^2\left(\vec{v}\times\vec{G}\right) = \vec{v}\times\left(\vec{\nabla}^2\vec{G}\right) - 2\,\vec{\Omega}\,\left(\vec{\nabla}\cdot\vec{G}\right)+2\,\vec{\nabla}\,\left(\vec{\Omega}\cdot\vec{G}\right). \tag{22}$$

Finally, taking into account that, to the first order in $\Omega$, $\vec{\Omega}\,\left(\vec{\nabla}\cdot\vec{G}\right)=0$, we get

$$\vec{\nabla}^2\left(\vec{v}\times\vec{G}\right) = \vec{v}\times\left(\vec{\nabla}^2\vec{G}\right)+2\,\vec{\nabla}\,\left(\vec{\Omega}\cdot\vec{G}\right). \tag{23}$$

Formulas (10)–(13) and (23) will be used in the next sections.

## 3. First system of wave equations for vectors $\vec{E}$ and $\vec{B}$ in a rotating medium

In this section, we are going to derive the first system of wave equations for electromagnetic field vectors $\vec{E}$ and $\vec{B}$ which will correspond to the original system of Maxwell's Eqs. (1–4).

### 3.1 Equation for vector $\vec{E}$

To derive the wave equation for vector $\vec{E}$, we apply the operator $\vec{\nabla}\times$ to expression (1):

$$\vec{\nabla}\times\left(\vec{\nabla}\times\vec{E}\right)+\frac{\partial}{\partial t}\left(\vec{\nabla}\times\vec{B}\right)=0. \tag{24}$$

Taking into account (10) and (3), we rewrite (24) in the form

$$\vec{\nabla}^2\,\vec{E}-\frac{n^2}{c^2}\frac{\partial^2\,\vec{E}}{\partial t^2}+\frac{1}{c^2}\frac{\partial}{\partial t}\left(\vec{v}\times\frac{\partial\,\vec{B}}{\partial t}\right)-\frac{1}{c^2}\frac{\partial}{\partial t}\left[\vec{\nabla}\times\left(\vec{v}\times\vec{E}\right)\right]-\vec{\nabla}\left(\vec{\nabla}\cdot\vec{E}\right)=0, \tag{25}$$

or, using (1),

$$\vec{\nabla}^2 \vec{E} - \frac{n^2}{c^2}\frac{\partial^2 \vec{E}}{\partial t^2} - \frac{1}{c^2}\frac{\partial}{\partial t}\left[\vec{\nabla} \times \left(\vec{v} \times \vec{E}\right) + \vec{v} \times \left(\vec{\nabla} \times \vec{E}\right)\right] - \vec{\nabla}\left(\vec{\nabla} \cdot \vec{E}\right) = 0. \quad (26)$$

With the help of (11), we get

$$\vec{\nabla}^2 \vec{E} - \frac{n^2}{c^2}\frac{\partial^2 \vec{E}}{\partial t^2} + \frac{1}{c^2}\frac{\partial}{\partial t}\left[\left(\vec{v} \cdot \vec{\nabla}\right)\vec{E} - \vec{\Omega} \times \vec{E} - \vec{v} \times \left(\vec{\nabla} \times \vec{E}\right)\right] - \vec{\nabla}\left(\vec{\nabla} \cdot \vec{E}\right) = 0. \quad (27)$$

Consider the term $\vec{\nabla}\left(\vec{\nabla} \cdot \vec{E}\right)$ in (27). According to (4), $\vec{\nabla} \cdot \vec{E} = \frac{1}{n^2}\vec{\nabla} \cdot \left(\vec{v} \times \vec{B}\right)$. Taking into account (12) and (3), we have $\vec{\nabla} \cdot \left(\vec{v} \times \vec{B}\right) = 2\,\vec{\Omega} \cdot \vec{B} - (n^2/c^2)(\partial/\partial t)\left(\vec{v} \cdot \vec{E}\right)$, so

$$\vec{\nabla}\left(\vec{\nabla} \cdot \vec{E}\right) = \frac{2}{n^2}\vec{\nabla}\left(\vec{\Omega} \cdot \vec{B}\right) - \frac{1}{c^2}\frac{\partial}{\partial t}\vec{\nabla}\left(\vec{v} \cdot \vec{E}\right). \quad (28)$$

Consider the last term in (28). In accordance with (13),

$$\vec{\nabla}\left(\vec{v} \cdot \vec{E}\right) = \left(\vec{v} \cdot \vec{\nabla}\right)\vec{E} - \vec{\Omega} \times \vec{E} + \vec{v} \times \left(\vec{\nabla} \times \vec{E}\right), \quad (29)$$

therefore

$$\vec{\nabla}\left(\vec{\nabla} \cdot \vec{E}\right) = \frac{2}{n^2}\vec{\nabla}\left(\vec{\Omega} \cdot \vec{B}\right) - \frac{1}{c^2}\frac{\partial}{\partial t}\left[\left(\vec{v} \cdot \vec{\nabla}\right)\vec{E} - \vec{\Omega} \times \vec{E} + \vec{v} \times \left(\vec{\nabla} \times \vec{E}\right)\right]. \quad (30)$$

Finally, substituting (30) into (27), we obtain the desired wave equation for vector $\vec{E}$:

$$\vec{\nabla}^2 \vec{E} - \frac{n^2}{c^2}\frac{\partial^2 \vec{E}}{\partial t^2} + \frac{2}{c^2}\frac{\partial}{\partial t}\left[\left(\vec{v} \cdot \vec{\nabla}\right)\vec{E} - \vec{\Omega} \times \vec{E}\right] - \frac{2}{n^2}\vec{\nabla}\left(\vec{\Omega} \cdot \vec{B}\right) = 0. \quad (31)$$

### 3.2 Equation for vector $\vec{B}$

In order to derive the wave equation for vector $\vec{B}$, we apply the operator $\vec{\nabla} \times$ to expression (3):

$$\vec{\nabla} \times \left(\vec{\nabla} \times \vec{B}\right) - \frac{1}{c^2}\vec{\nabla} \times \left[\vec{\nabla} \times \left(\vec{v} \times E\right)\right] - \frac{n^2}{c^2}\frac{\partial}{\partial t}\left(\vec{\nabla} \times \vec{E}\right) + \frac{1}{c^2}\frac{\partial}{\partial t}\left[\vec{\nabla} \times \left(\vec{v} \times \vec{B}\right)\right] = 0. \quad (32)$$

Taking into account (10), (1), and (2), we rewrite (32) as

$$\vec{\nabla}^2 \vec{B} - \frac{n^2}{c^2}\frac{\partial^2 \vec{B}}{\partial t^2} - \frac{1}{c^2}\frac{\partial}{\partial t}\left[\vec{\nabla} \times \left(\vec{v} \times \vec{B}\right)\right] + \frac{1}{c^2}\vec{\nabla} \times \left[\vec{\nabla} \times \left(\vec{v} \times \vec{E}\right)\right] = 0. \quad (33)$$

Since $\vec{\nabla} \times \left[\vec{\nabla} \times \left(\vec{v} \times \vec{E}\right)\right] = -\vec{\nabla}^2\left(\vec{v} \times \vec{E}\right) + \vec{\nabla}\left[\vec{\nabla} \cdot \left(\vec{v} \times \vec{E}\right)\right]$,

$$\vec{\nabla}^2 \vec{B} - \frac{n^2}{c^2}\frac{\partial^2 \vec{B}}{\partial t^2} - \frac{1}{c^2}\frac{\partial}{\partial t}\left[\vec{\nabla} \times \left(\vec{v} \times \vec{B}\right)\right] - \frac{1}{c^2}\vec{\nabla}^2\left(\vec{v} \times \vec{E}\right) + \frac{1}{c^2}\vec{\nabla}\left[\vec{\nabla}\cdot\left(\vec{v} \times \vec{E}\right)\right] = 0. \tag{34}$$

Using (11), we have

$$\vec{\nabla}^2 \vec{B} - \frac{n^2}{c^2}\frac{\partial^2 \vec{B}}{\partial t^2} + \frac{1}{c^2}\frac{\partial}{\partial t}\left[\left(\vec{v}\cdot\vec{\nabla}\right)\vec{B} - \vec{\Omega} \times \vec{B}\right] - \frac{1}{c^2}\vec{\nabla}^2\left(\vec{v} \times \vec{E}\right) + \frac{1}{c^2}\vec{\nabla}\left[\vec{\nabla}\cdot\left(\vec{v} \times \vec{E}\right)\right] = 0. \tag{35}$$

Consider the term $\vec{\nabla}\left[\vec{\nabla}\cdot\left(\vec{v} \times \vec{E}\right)\right]$ in (35). In accordance with (12), $\vec{\nabla}\cdot\left(\vec{v} \times \vec{E}\right) = 2\,\vec{\Omega}\cdot\vec{E} - \vec{v}\cdot\left(\vec{\nabla} \times \vec{E}\right)$, or, with (1), $\vec{\nabla}\cdot\left(\vec{v} \times \vec{E}\right) = 2\,\vec{\Omega}\cdot\vec{E} + (\partial/\partial t)\left(\vec{v}\cdot\vec{B}\right)$. Then

$$\vec{\nabla}\left[\vec{\nabla}\cdot\left(\vec{v} \times \vec{E}\right)\right] = 2\,\vec{\nabla}\left(\vec{\Omega}\cdot\vec{E}\right) + \frac{\partial}{\partial t}\vec{\nabla}\left(\vec{v}\cdot\vec{B}\right), \tag{36}$$

or, taking into account (13),

$$\vec{\nabla}\left[\vec{\nabla}\cdot\left(\vec{v} \times \vec{E}\right)\right] = 2\,\vec{\nabla}\left(\vec{\Omega}\cdot\vec{E}\right) + \frac{\partial}{\partial t}\left[\left(\vec{v}\cdot\vec{\nabla}\right)\vec{B} - \vec{\Omega} \times \vec{B}\right] + \frac{\partial}{\partial t}\left[\vec{v} \times \left(\vec{\nabla} \times \vec{B}\right)\right]. \tag{37}$$

With the help of (3), we find

$$\vec{\nabla}\left[\vec{\nabla}\cdot\left(\vec{v} \times \vec{E}\right)\right] = 2\,\vec{\nabla}\left(\vec{\Omega}\cdot\vec{E}\right) + \frac{\partial}{\partial t}\left[\left(\vec{v}\cdot\vec{\nabla}\right)\vec{B} - \vec{\Omega} \times \vec{B}\right] + \vec{v} \times \frac{n^2}{c^2}\frac{\partial^2 \vec{E}}{\partial t^2}. \tag{38}$$

Substituting (38) into (35), we obtain

$$\vec{\nabla}^2 \vec{B} - \frac{n^2}{c^2}\frac{\partial^2 \vec{B}}{\partial t^2} + \frac{2}{c^2}\frac{\partial}{\partial t}\left[\left(\vec{v}\cdot\vec{\nabla}\right)\vec{B} - \vec{\Omega} \times \vec{B}\right] + \frac{2}{c^2}\vec{\nabla}\left(\vec{\Omega}\cdot\vec{E}\right) - \frac{1}{c^2}\left[\vec{\nabla}^2\left(\vec{v} \times \vec{E}\right) - \vec{v} \times \frac{n^2}{c^2}\frac{\partial^2 \vec{E}}{\partial t^2}\right] = 0. \tag{39}$$

Consider the last term in (39). In accordance with (23), $\vec{\nabla}^2\left(\vec{v} \times \vec{E}\right) = \vec{v} \times \left(\vec{\nabla}^2 \vec{E}\right) + 2\,\vec{\nabla}\left(\vec{\Omega}\cdot\vec{E}\right)$; hence

$$\vec{\nabla}^2\left(\vec{v} \times \vec{E}\right) - \vec{v} \times \frac{n^2}{c^2}\frac{\partial^2 \vec{E}}{\partial t^2} = \vec{v} \times \left(\vec{\nabla}^2 \vec{E} - \frac{n^2}{c^2}\frac{\partial^2 \vec{E}}{\partial t^2}\right) + 2\,\vec{\nabla}\left(\vec{\Omega}\cdot\vec{E}\right). \tag{40}$$

It is clear that, to the first order in $\Omega$,

$$\vec{v} \times \left(\vec{\nabla}^2 \vec{E} - \frac{n^2}{c^2}\frac{\partial^2 \vec{E}}{\partial t^2}\right) = 0. \tag{41}$$

So, expression (40) may be rewritten as

$$\vec{\nabla}^2\left(\vec{v} \times \vec{E}\right) - \vec{v} \times \frac{n^2}{c^2}\frac{\partial^2 \vec{E}}{\partial t^2} = 2\,\vec{\nabla}\left(\vec{\Omega}\cdot\vec{E}\right). \tag{42}$$

After inserting (42) into (39), we obtain the desired wave equation for vector $\vec{B}$:

$$\vec{\nabla}^2\vec{B} - \frac{n^2}{c^2}\frac{\partial^2\vec{B}}{\partial t^2} + \frac{2}{c^2}\frac{\partial}{\partial t}\left[\left(\vec{v}\cdot\vec{\nabla}\right)\vec{B} - \vec{\Omega}\times\vec{B}\right] = 0. \tag{43}$$

### 3.3 Result of Section 3

According to (31) and (43), the first system of wave equations for electromagnetic field vectors $\vec{E}$ and $\vec{B}$ in a rotating medium has the following form:

$$\begin{aligned}
&\vec{\nabla}^2\vec{E} - \frac{n^2}{c^2}\frac{\partial^2\vec{E}}{\partial t^2} + \frac{2}{c^2}\frac{\partial}{\partial t}\left[\left(\vec{v}\cdot\vec{\nabla}\right)\vec{E} - \vec{\Omega}\times\vec{E}\right] - \frac{2}{n^2}\vec{\nabla}\left(\vec{\Omega}\cdot\vec{B}\right) = 0,\\
&\vec{\nabla}^2\vec{B} - \frac{n^2}{c^2}\frac{\partial^2\vec{B}}{\partial t^2} + \frac{2}{c^2}\frac{\partial}{\partial t}\left[\left(\vec{v}\cdot\vec{\nabla}\right)\vec{B} - \vec{\Omega}\times\vec{B}\right] = 0.
\end{aligned} \tag{44}$$

## 4. Second system of wave equations for vectors $\vec{E}$ and $\vec{B}$ in a rotating medium

In this section, we are going to obtain the second system of wave equations for electromagnetic field vectors $\vec{E}$ and $\vec{B}$ which will correspond to the original system of Maxwell's Eqs. (5–8).

### 4.1 Equation for vector $\vec{E}$

To derive the wave equation for vector $\vec{E}$, we apply the operator $\vec{\nabla}\times$ to expression (5):

$$\vec{\nabla}\times\left(\vec{\nabla}\times\vec{E}\right) + \frac{\partial}{\partial t}\left(\vec{\nabla}\times\vec{B}\right) + \frac{1}{c^2}\frac{\partial}{\partial t}\left[\vec{\nabla}\times\left(\vec{v}\times\vec{E}\right)\right] = 0. \tag{45}$$

Taking into consideration (7) and (10), we rewrite (45) in the form

$$\vec{\nabla}^2\vec{E} - \frac{n^2}{c^2}\frac{\partial^2\vec{E}}{\partial t^2} + \frac{1}{c^2}\frac{\partial}{\partial t}\left(\vec{v}\times\frac{\partial\vec{B}}{\partial t}\right) - \frac{1}{c^2}\frac{\partial}{\partial t}\left[\vec{\nabla}\times\left(\vec{v}\times\vec{E}\right)\right] - \vec{\nabla}\left(\vec{\nabla}\cdot\vec{E}\right) = 0, \tag{46}$$

or, using (5),

$$\vec{\nabla}^2\vec{E} - \frac{n^2}{c^2}\frac{\partial^2\vec{E}}{\partial t^2} - \frac{1}{c^2}\frac{\partial}{\partial t}\left[\vec{\nabla}\times\left(\vec{v}\times\vec{E}\right) + \vec{v}\times\left(\vec{\nabla}\times\vec{E}\right)\right] - \vec{\nabla}\left(\vec{\nabla}\cdot\vec{E}\right) = 0. \tag{47}$$

Taking into account (11), we have

$$\vec{\nabla}^2\vec{E} - \frac{n^2}{c^2}\frac{\partial^2\vec{E}}{\partial t^2} + \frac{1}{c^2}\frac{\partial}{\partial t}\left[\left(\vec{v}\cdot\vec{\nabla}\right)\vec{E} - \vec{\Omega}\times\vec{E} - \vec{v}\times\left(\vec{\nabla}\times\vec{E}\right)\right] - \vec{\nabla}\left(\vec{\nabla}\cdot\vec{E}\right) = 0. \tag{48}$$

Consider the term $\vec{\nabla}\left(\vec{\nabla}\cdot\vec{E}\right)$ in (48). According to (8), $\vec{\nabla}\cdot\vec{E}=\frac{1}{n^2}\,\vec{\nabla}\cdot\left(\vec{v}\times\vec{B}\right)$, where, taking into account (12),

$$\vec{\nabla}\cdot\left(\vec{v}\times\vec{B}\right)=2\,\vec{\Omega}\cdot\vec{B}-\vec{v}\cdot\left(\vec{\nabla}\times\vec{B}\right). \tag{49}$$

With the help of (7), we find $\vec{\nabla}\cdot\left(\vec{v}\times\vec{B}\right)=2\,\vec{\Omega}\cdot\vec{B}-(n^2/c^2)\,(\partial/\partial t)\left(\vec{v}\cdot\vec{E}\right)$; hence

$$\vec{\nabla}\left(\vec{\nabla}\cdot\vec{E}\right)=\frac{2}{n^2}\,\vec{\nabla}\left(\vec{\Omega}\cdot\vec{B}\right)-\frac{1}{c^2}\frac{\partial}{\partial t}\,\vec{\nabla}\left(\vec{v}\cdot\vec{E}\right). \tag{50}$$

Consider the term $\vec{\nabla}\left(\vec{v}\cdot\vec{E}\right)$ in (50). In accordance with (13),

$$\vec{\nabla}\left(\vec{v}\cdot\vec{E}\right)=\left(\vec{v}\cdot\vec{\nabla}\right)\vec{E}-\vec{\Omega}\times\vec{E}+\vec{v}\times\left(\vec{\nabla}\times\vec{E}\right). \tag{51}$$

Then

$$\vec{\nabla}\left(\vec{\nabla}\cdot\vec{E}\right)=\frac{2}{n^2}\,\vec{\nabla}\left(\vec{\Omega}\cdot\vec{B}\right)-\frac{1}{c^2}\frac{\partial}{\partial t}\left[\left(\vec{v}\cdot\vec{\nabla}\right)\vec{E}-\vec{\Omega}\times\vec{E}+\vec{v}\times\left(\vec{\nabla}\times\vec{E}\right)\right]. \tag{52}$$

Substituting (52) into (48), we obtain the desired wave equation for vector $\vec{E}$:

$$\vec{\nabla}^2\,\vec{E}-\frac{n^2}{c^2}\frac{\partial^2\,\vec{E}}{\partial t^2}+\frac{2}{c^2}\frac{\partial}{\partial t}\left[\left(\vec{v}\cdot\vec{\nabla}\right)\vec{E}-\vec{\Omega}\times\vec{E}\right]-\frac{2}{n^2}\,\vec{\nabla}\left(\vec{\Omega}\cdot\vec{B}\right)=0. \tag{53}$$

### 4.2 Equation for vector $\vec{B}$

In order to derive the wave equation for vector $\vec{B}$, we apply the operator $\vec{\nabla}\times$ to expression (7):

$$\vec{\nabla}\times\left(\vec{\nabla}\times\vec{B}\right)-\frac{n^2}{c^2}\frac{\partial}{\partial t}\left(\vec{\nabla}\times\vec{E}\right)+\frac{1}{c^2}\frac{\partial}{\partial t}\left[\vec{\nabla}\times\left(\vec{v}\times\vec{B}\right)\right]=0. \tag{54}$$

Taking into account (5) and (10), we rewrite (54) in the form

$$\vec{\nabla}^2\,\vec{B}-\frac{n^2}{c^2}\frac{\partial^2\,\vec{B}}{\partial t^2}-\frac{1}{c^2}\frac{\partial}{\partial t}\left(\vec{v}\times\frac{n^2}{c^2}\frac{\partial\,\vec{E}}{\partial t}\right)-\frac{1}{c^2}\frac{\partial}{\partial t}\left[\vec{\nabla}\times\left(\vec{v}\times\vec{B}\right)\right]-\vec{\nabla}\left(\vec{\nabla}\cdot\vec{B}\right)=0, \tag{55}$$

or, using (7),

$$\vec{\nabla}^2\,\vec{B}-\frac{n^2}{c^2}\frac{\partial^2\,\vec{B}}{\partial t^2}-\frac{1}{c^2}\frac{\partial}{\partial t}\left[\vec{\nabla}\times\left(\vec{v}\times\vec{B}\right)+\vec{v}\times\left(\vec{\nabla}\times\vec{B}\right)\right]-\vec{\nabla}\left(\vec{\nabla}\cdot\vec{B}\right)=0. \tag{56}$$

Consider the term $\vec{\nabla}\times\left(\vec{v}\times\vec{B}\right)$ in (56). According to (11), $\vec{\nabla}\times\left(\vec{v}\times\vec{B}\right)=-\left(\vec{v}\cdot\vec{\nabla}\right)\vec{B}+\vec{\Omega}\times\vec{B}$. Thus,

$$\vec{\nabla}^2\vec{B} - \frac{n^2}{c^2}\frac{\partial^2\vec{B}}{\partial t^2} + \frac{1}{c^2}\frac{\partial}{\partial t}\left[\left(\vec{v}\cdot\vec{\nabla}\right)\vec{B} - \vec{\Omega}\times\vec{B} - \vec{v}\times\left(\vec{\nabla}\times\vec{B}\right)\right] - \vec{\nabla}\left(\vec{\nabla}\cdot\vec{B}\right) = 0. \tag{57}$$

Consider the term $\vec{\nabla}\left(\vec{\nabla}\cdot\vec{B}\right)$ in (57). In accordance with (6),

$$\vec{\nabla}\cdot\vec{B} = -\left(1/c^2\right)\vec{\nabla}\cdot\left(\vec{v}\times\vec{E}\right). \tag{58}$$

Taking into account (12), we have

$$\vec{\nabla}\cdot\left(\vec{v}\times\vec{E}\right) = 2\,\vec{\Omega}\cdot\vec{E} - \vec{v}\cdot\left(\vec{\nabla}\times\vec{E}\right), \tag{59}$$

or, using (5),

$$\vec{\nabla}\cdot\left(\vec{v}\times\vec{E}\right) = 2\,\vec{\Omega}\cdot\vec{E} + \frac{\partial}{\partial t}\left(\vec{v}\cdot\vec{B}\right). \tag{60}$$

So

$$\vec{\nabla}\cdot\vec{B} = -\frac{2}{c^2}\,\vec{\Omega}\cdot\vec{E} - \frac{1}{c^2}\frac{\partial}{\partial t}\left(\vec{v}\cdot\vec{B}\right). \tag{61}$$

Hence

$$\vec{\nabla}\left(\vec{\nabla}\cdot\vec{B}\right) = -\frac{2}{c^2}\,\vec{\nabla}\left(\vec{\Omega}\cdot\vec{E}\right) - \frac{1}{c^2}\frac{\partial}{\partial t}\,\vec{\nabla}\left(\vec{v}\cdot\vec{B}\right). \tag{62}$$

Consider the term $\vec{\nabla}\left(\vec{v}\cdot\vec{B}\right)$ in (62). In accordance with (13),

$$\vec{\nabla}\left(\vec{v}\cdot\vec{B}\right) = \left(\vec{v}\cdot\vec{\nabla}\right)\vec{B} - \vec{\Omega}\times\vec{B} + \vec{v}\times\left(\vec{\nabla}\times\vec{B}\right). \tag{63}$$

Then

$$\vec{\nabla}\left(\vec{\nabla}\cdot\vec{B}\right) = -\frac{2}{c^2}\,\vec{\nabla}\left(\vec{\Omega}\cdot\vec{E}\right) - \frac{1}{c^2}\frac{\partial}{\partial t}\left[\left(\vec{v}\cdot\vec{\nabla}\right)\vec{B} - \vec{\Omega}\times\vec{B} + \vec{v}\times\left(\vec{\nabla}\times\vec{B}\right)\right]. \tag{64}$$

Substituting (64) into (57), we obtain the desired wave equation for vector $\vec{B}$:

$$\vec{\nabla}^2\vec{B} - \frac{n^2}{c^2}\frac{\partial^2\vec{B}}{\partial t^2} + \frac{2}{c^2}\frac{\partial}{\partial t}\left[\left(\vec{v}\cdot\vec{\nabla}\right)\vec{B} - \vec{\Omega}\times\vec{B}\right] + \frac{2}{c^2}\,\vec{\nabla}\left(\vec{\Omega}\cdot\vec{E}\right) = 0. \tag{65}$$

### 4.3 Result of Section 4

According to (53) and (65), the second system of wave equations for electromagnetic field vectors $\vec{E}$ and $\vec{B}$ in a rotating medium has the following form:

$$\begin{aligned}
&\vec{\nabla}^2\vec{E} - \frac{n^2}{c^2}\frac{\partial^2\vec{E}}{\partial t^2} + \frac{2}{c^2}\frac{\partial}{\partial t}\left[\left(\vec{v}\cdot\vec{\nabla}\right)\vec{E} - \vec{\Omega}\times\vec{E}\right] - \frac{2}{n^2}\vec{\nabla}\left(\vec{\Omega}\cdot\vec{B}\right) = 0,\\
&\vec{\nabla}^2\vec{B} - \frac{n^2}{c^2}\frac{\partial^2\vec{B}}{\partial t^2} + \frac{2}{c^2}\frac{\partial}{\partial t}\left[\left(\vec{v}\cdot\vec{\nabla}\right)\vec{B} - \vec{\Omega}\times\vec{B}\right] + \frac{2}{c^2}\vec{\nabla}\left(\vec{\Omega}\cdot\vec{E}\right) = 0.
\end{aligned} \tag{66}$$

## 5. Two systems of wave equations for vectors $\vec{E}$ and $\vec{B}$ in a rotating medium: a comparative analysis

The first system of wave equations for electromagnetic field vectors $\vec{E}$ and $\vec{B}$ [which was derived from the corresponding original system of Maxwell's Eqs. (1–4)] has the form

$$\vec{\nabla}^2 \vec{E} - \frac{n^2}{c^2}\frac{\partial^2 \vec{E}}{\partial t^2} + \frac{2}{c^2}\frac{\partial}{\partial t}\left[\left(\vec{v}\cdot\vec{\nabla}\right)\vec{E} - \vec{\Omega}\times\vec{E}\right] - \frac{2}{n^2}\vec{\nabla}\left(\vec{\Omega}\cdot\vec{B}\right) = 0, \quad (67)$$

$$\vec{\nabla}^2 \vec{B} - \frac{n^2}{c^2}\frac{\partial^2 \vec{B}}{\partial t^2} + \frac{2}{c^2}\frac{\partial}{\partial t}\left[\left(\vec{v}\cdot\vec{\nabla}\right)\vec{B} - \vec{\Omega}\times\vec{B}\right] = 0. \quad (68)$$

The second system of wave equations [which was obtained from the corresponding original system of Maxwell's Eqs. (5–8)] is

$$\vec{\nabla}^2 \vec{E} - \frac{n^2}{c^2}\frac{\partial^2 \vec{E}}{\partial t^2} + \frac{2}{c^2}\frac{\partial}{\partial t}\left[\left(\vec{v}\cdot\vec{\nabla}\right)\vec{E} - \vec{\Omega}\times\vec{E}\right] - \frac{2}{n^2}\vec{\nabla}\left(\vec{\Omega}\cdot\vec{B}\right) = 0, \quad (69)$$

$$\vec{\nabla}^2 \vec{B} - \frac{n^2}{c^2}\frac{\partial^2 \vec{B}}{\partial t^2} + \frac{2}{c^2}\frac{\partial}{\partial t}\left[\left(\vec{v}\cdot\vec{\nabla}\right)\vec{B} - \vec{\Omega}\times\vec{B}\right] + \frac{2}{c^2}\vec{\nabla}\left(\vec{\Omega}\cdot\vec{E}\right) = 0. \quad (70)$$

Systems (67–70) have been derived in preprint [14] on the base of the procedure developed (for the case of vacuum) in work [15]. It should be noted that Eqs. (67) and (69) for vector $\vec{E}$ in both systems, after dropping the terms $\vec{\Omega}\times\vec{E}$ and $\vec{\nabla}\left(\vec{\Omega}\cdot\vec{B}\right)$, take the approximate form

$$\vec{\nabla}^2 \vec{E} - \frac{n^2}{c^2}\frac{\partial^2 \vec{E}}{\partial t^2} + \frac{2}{c^2}\frac{\partial}{\partial t}\left[\left(\vec{v}\cdot\vec{\nabla}\right)\vec{E}\right] = 0, \quad (71)$$

known, for example, from work [8] (see relation (33) therein).

Expressions (67, 68) and (69, 70) represent two qualitatively different systems of wave equations for vectors $\vec{E}$ and $\vec{B}$ in a uniformly rotating dielectric medium. From a comparative analysis of these systems, it follows:

1. The structure of equations for vectors $\vec{E}$ and $\vec{B}$ in the first system (67, 68) is asymmetrical with respect to $\Omega$. Therefore, the propagation of $\vec{E}$- and $\vec{B}$-components of electromagnetic waves in a rotating medium in general case $\vec{\Omega}= \Omega_x\hat{x} + \Omega_y\hat{y} + \Omega_z\hat{z}$ will be governed by qualitatively different laws.

2. The structure of the wave equations in the second system (69, 70) is symmetrical with respect to $\Omega$. Hence, the propagation of the named field components in such case will be governed by similar laws.

Systems of wave Eqs. (67, 68) and (69, 70) [in conjunction with corresponding original sets of Maxwell's Eqs. (1–4) and (5–8)] may serve as a theoretical basis for a detailed study of the process of electromagnetic wave propagation in a rotating dielectric medium. But before the beginning of such study, the researcher must first solve the problem of choosing between them (because the final results will be different).

[In the author's opinion, from the position of the principle of symmetry, the system of wave Eqs. (69, 70) is preferable to system (67, 68), and the corresponding original system of Maxwell's Eqs. (5–8) is preferable to the original system (1–4).]

## 6. Conclusion

In this chapter, on the base of two basic systems of Maxwell's Eqs. (1–4) and (5–8) for electromagnetic field vectors $\vec{E}$ and $\vec{B}$ in a uniformly rotating dielectric medium, the two corresponding systems of wave Eqs. (67, 68) and (69, 70) have been derived (to the first order in $\Omega$).

From their comparative analysis, it can be seen that the structure of the wave equations for electromagnetic field vectors in system (67, 68) is asymmetrical with respect to $\Omega$, while the structure of such equations for these vectors in system (69, 70) is symmetrical.

On this basis, it can be concluded that if the principle of symmetry is accepted as a criterion for selection, then system of wave Eqs. (69, 70) for electromagnetic field vectors $\vec{E}$ and $\vec{B}$ may be preferred. The same conclusion may be also made about the original system of Maxwell's Eqs. (5–8) on the base of which set of wave Eqs. (69, 70) has been obtained.

## Author details

Evgeny A. Bondarenko
National Technical University of Ukraine, 'Igor Sikorsky Kyiv Polytechnic Institute', Kyiv, Ukraine

*Address all correspondence to: ea_bndrk@ukr.net

## References

[1] Heer CV. Resonant frequencies of an electromagnetic cavity in an accelerated system of reference. Physical Review. 1964;**134**:A799-A804

[2] Mo TC. Theory of electrodynamics in media in noninertial frames and applications. Journal of Mathematical Physics. 1970;**11**:2589-2610

[3] Khromykh AM. Ring generator in a rotating reference system. Soviet Physics JETP. 1966;**23**:185-186

[4] Anderson JL, Ryon JW. Electromagnetic radiation in accelerated systems. Physical Review. 1969;**181**:1765-1775

[5] Lianis G, Whicker D. Electromagnetic phenomena in rotating media. Archive for Rational Mechanics and Analysis. 1975;**57**:325-362

[6] Korostelev AA, Fateev VF. Electrodynamics of moving media in noninertial reference systems as applied to the processes in a ring resonator. Optika i Spektroskopiia. 1978;**45**:132-139. In Russian

[7] Shpak IV, Solomin AV. Effect of the refractive index of the medium on the splitting of the eigenfrequencies of a rotating ring cavity. Optics and Spectroscopy. 1979;**46**:72-75

[8] Scorgie GC. Electromagnetism in non-inertial coordinates. Journal of Physics A: Mathematical and General. 1990;**23**:5169-5184

[9] Milovskii ND. Equations of electrodynamics in a rotating solid dielectric. Optics and Spectroscopy. 2017;**123**:642-649

[10] Schiff LI. A question in general relativity. Proceedings of the National Academy of Sciences of the United States of America. 1939;**25**:391-395

[11] Irvine WM. Electrodynamics in a rotating system of reference. Physica. 1964;**30**:1160-1170

[12] Landau LD, Lifshitz EM. The Classical Theory of Fields. New York: Pergamon Press; 1971

[13] Korn GA, Korn TM. Mathematical Handbook for Scientists and Engineers. New York: McGraw-Hill Book Company; 1968

[14] Bondarenko EA. First-order derivation of two basic systems of wave equations for electromagnetic field vectors $\vec{E}$ and $\vec{B}$ in a uniformly rotating dielectric medium. ResearchGate. 2018. Preprint. DOI: 10.13140/RG.2.2.17459.94241

[15] Bondarenko EA. Two systems of Maxwell's equations and two corresponding systems of wave equations for electromagnetic field vectors $\vec{E}$ and $\vec{B}$ in a rotating frame of reference: A linear approximation. Ukrainian Journal of Physics. 2017;**62**:99-105

Chapter 6

# The Interaction of Microwaves with Materials of Different Properties

*Rafael Zamorano Ulloa,*
*Ma. Guadalupe Hernandez Santiago*
*and Veronica L. Villegas Rueda*

## Abstract

Electromagnetic radiation, such as microwaves, are all the time reflected, transmitted, and/or absorbed by any kind of matter, glasses, conductors, water, ferrites, and so forth. Magnetic materials absorb greatly microwaves. The more magnetic, the more microwaves are absorbed. The aim of this chapter is to present the fundamental physics of the absorption of microwave power (energy per unit time) by ferrimagnetic and ferromagnetic matter in the nano and micro size scale. The magnetic moments and their collective modes are the basic microscopic absorbers under in-resonance and out-of-resonance conditions. Experimental setups and measurement techniques are described. The profiles of microwave absorption are described and connected to the micromagnetic environment that elicits such absorption. Section by section and the overall microwave power absorption profiles are related to the micromagnetic structures. Emphasis is made on nano- and micromagnets. These interactions of microwaves with nano- and micromagnets serve to infer microscopic magnetic information.

**Keywords:** microwaves technology, microwaves absorption, magnetic materials, reflection and transmission microwaves

## 1. Introduction

Microwaves pervade the whole universe; this was discovered by quite an accident by R. W. Wilson and A. A. Penzias in 1964 [1–3]; microwaves seem to bathe the earth and the space from all directions. Microwaves, also, pervade today's world of high technology [4, 5]. They are of tremendous scientific interest, and they are, just, indispensable in today's communication [6], military [7], medical [8], domestic appliances, scientific instruments [9, 10] (electron spin resonance, ESR/ferromagnetic resonance spectroscopy, FMR), microwave passive instruments [11–15], radars, spaceships, satellites, and so forth; they are also found in many industries such as automobile, data, memory and computer processing, and microwave instrumentation [14, 15]; the signal processing in the range of 5–50 GHz is quite interesting for security, military, and communication applications [15, 16]. Modern microwave communication especially mobile communication and satellites requires high performance band-stop filter having high frequency selectivity, smaller size, high stop band attenuation, and low insertion loss [15, 17]. A good account of different applications is found in [17].

Most impressive, they are already playing an important role in the development of smart cities along with smart transportation, smart energy, smart health care, and so forth [18]. **Figure 1** shows an electromagnetic spectrum, making emphasis on the microwave region and the definition of its different bands. **Figure 2** shows an assortment of pictorial representations of some of the modern applications of microwaves.

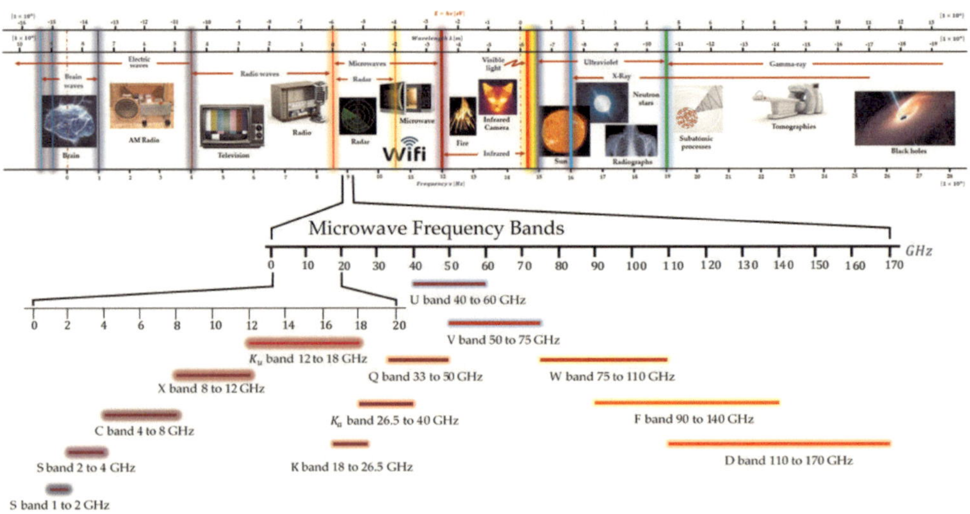


**Figure 1.**
*The electromagnetic spectrum, making emphasis on the microwave region and its different bands.*

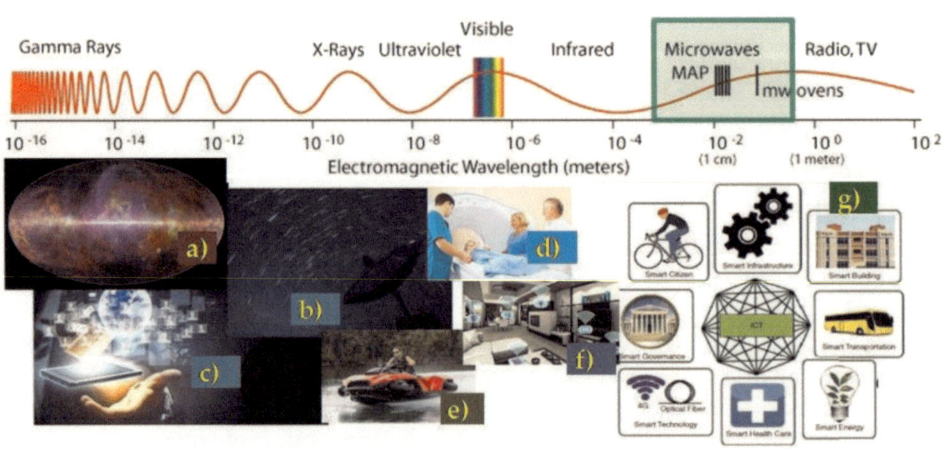


**Figure 2.**
*The microwaves in the electromagnetic spectrum range from 300 MHz (1 m) to 300 GHz (1 mm) [3]. There are microwaves present in: (a) and (b) the whole universe, a microwave view captured by the European Space Agency's Planck satellite [19]; (c) the digital technology [20, 21]; (d) biomedical use [22]; (e) controversial proposition to propel starships [23]; (f) in telecommunications, Wi-Fi antennas; and (g) in the development of smart cities [18].*

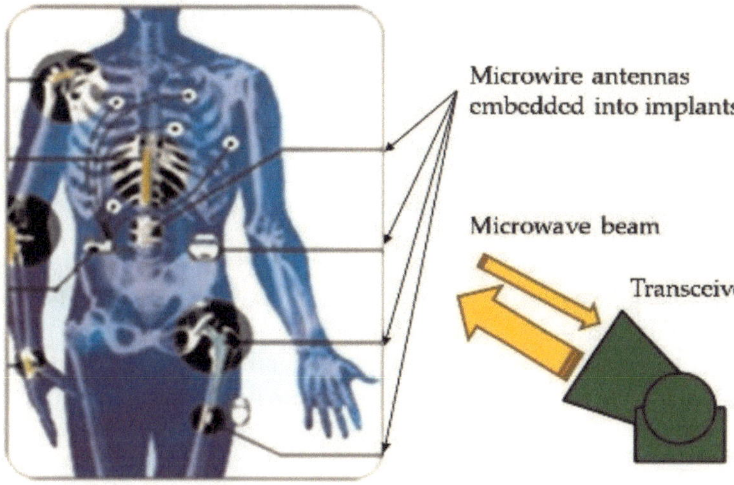


**Figure 3.**
*Microwaves as the signals coming in and out from magnetic microwire sensors implanted in injured-surgery patients in order to follow postoperatory recovery [8]. The transceiver houses, both, the microwave source and the microwave detectors. Probably with the capacity to detect anisotropic distribution of signals.*

In **Figure 3**, we show one of the most impactful applications of microwaves as the signals coming in and out from magnetic microwire sensors implanted in injured-surgery patients in order to follow postoperatory recovery [8]. The microwave beam should penetrate a considerable section of tissue then "hit" the magnetic wire, film, or rod, reflecting in many directions. Some microwaves will get back to the horn entrance of the transceiver, then detected and processed. The technology already exists and is already in use in atmospheric sciences [21] and in gun radars to detect speedy objects. These applications cannot be carried out with laser, for example, as laser light cannot penetrate tissue as microwaves do. It should be noted that in all the above examples, microwaves and radar work in open spaces and "hit" a target, and then some of the reflected beams are detected. Absorption in the medium itself and by obstacles reduces the reading at the detectors. There are also a great number of uses of microwaves inside tubes, pipes, and cavities. More controlled energy flux is attained and precision measurements can be done. We treat them after we give the fundamentals of the physics involved.

But, why are they (microwave) everywhere in this world and out of this world? It is because of the peculiarity of their electromagnetic properties and the way they interact with matter. They started to propagate through the universe at the epoch of recombination in the cosmic evolution. The cosmic microwave background (CMB) is the most ancient relic we have today of the beginning of the universe in the form of a big bang [24, 25]. It is the oldest electromagnetic radiation, older than visible light [26]. When atoms and nuclei appeared for the first time in the universe, microwaves started to interact with them immediately. Microwaves interact with atoms, nuclei, protons, electrons, molecules, clusters of molecules, and so forth. At the macroscopic scale, microwaves interact with all kinds of matter: rocks, gases, clouds, liquids (water and oceans), dielectrics, plasmas, ionosphere, metals, magnetic matter, and so on. It gets reflected, transmitted, and very frequently absorbed. Microwaves make atoms rotate (rotational excitation) and make electric dipoles jiggle frenetically, and when electric dipoles are part of dielectric materials, microwaves heat them. Microwaves make magnetic dipoles rotate and jump up magnetic energy states. The free electrons in metallic objects absorb greatly microwaves.

## 2. The electrodynamic properties of matter interact strongly with microwaves

Microwaves interact with matter, microscopically, through its constituent atoms, conduction electrons if present, and atomic magnetic dipoles if present. Yet, macroscopically, the effects of microwaves on matter are well described by the four Maxwell equations and the electrodynamic properties of matter: $\epsilon$ (electric permittivity), $\mu$ (magnetic permeability), and $\sigma$ (electrical conductivity).

The interactions of microwaves with matter are of many kinds. The general electrodynamic properties of matter, $\epsilon$, $\mu$, and $\sigma$, determine completely their behavior when microwaves "hit" them. More specifically, the electric permittivity, $\epsilon$, carries information on the polarization of a dielectric specimen (water, vapor, clouds, wood, glass, and so on) and is related to the number of electric dipoles as $\chi = N\alpha/(\epsilon_0 - N\alpha b)$ and $P = (\epsilon - \epsilon_0)\ E$, and $\epsilon = (1 + \chi)\ \epsilon_0$, with $P = \epsilon_0 \chi E$ and $\alpha$ the molecular polarizability of the medium and is generally anisotropic, i.e., $\alpha_x \neq \alpha_y \neq \alpha_z$; hence $\chi$ in general is anisotropic and is represented by a tensor in matrix form. Electric dipoles absorb greatly microwaves because these cause the electric dipoles to execute damped oscillations at the GHz frequency. The damped motion brings with it a complex $\epsilon = \epsilon' - i\,\epsilon''$ which is also a function of frequency [27, 28], in which $\epsilon''$ takes account of the energy losses.

The magnetic permeability, $\mu$, carries information on the magnetization capacity of a material that carries a number N of magnetic dipoles. They are related by $\mu = \mu_0(1 + \chi_m)$, $\mathbf{M} = \chi_m\mathbf{H}$, $\mathbf{M} = (\mu_r - 1)\mathbf{H}$, and $\mathbf{M} = \sum\mathbf{m}_i$, where $\mathbf{m}_i$ are microscopic, atomic magnetic moments (spin, S; orbital, m) [29, 30]. Magnetic dipoles absorb microwave energy because they precess with damping under the torques produced by the microwave's magnetic field; according to the Landau-Lifshitz [30] equation of motion: $\mathbf{M}'(t) = \gamma\mathbf{M}\times\mathbf{H}(\omega) - \alpha\mathbf{M}\times(\mathbf{M}\times\mathbf{H}(\omega))$, in which $\mathbf{H}(\omega)$ is the magnetic field component of the microwaves, $\gamma$ is the gyromagnetic ratio, and $\alpha$ is the damping constant. The precession velocity and hence $\mathbf{M}'$ is different for different $\mathbf{H}(\omega)$. The higher the frequency, the higher the losses. The damped precessions bring with them the loss of microwave energy making the magnetic permeability complex and frequency-dependent, $\mu(\omega) = \mu'(\omega) - i\mu''(\omega)$. In addition, the response of $\mathbf{M}$ to $\mathbf{H}(\omega)$ is almost always direction-dependent, i.e., given H in direction x produces $M_x$, but the same H applied along y or z produces $M_y \neq M_z \neq M_x$, and this responses are properly described with a tensor $\boldsymbol{\chi}_\mathbf{m}(\omega)$, or tensor $\mu(\omega)$. When the magnetic material is ferro- or ferrimagnetic and it is not magnetically saturated, its magnetic structure is comprised of domains and domain walls; the magnetization, $M_a$, within a domain, a, has a magnitude and a direction, **a**; the magnetization, $M_b$, within domain b, has another magnitude and another direction, **b**, and so on. The walls in between the domains have a considerable amount of magnetic energy [30, 31] and can move in translational or rotating dissipative and anisotropic motion following the LL damped equation of motion given above. An iconic set of magnetic materials that has been used in multiple microwave applications since its invention is the so-called microwave ferrites [11–15]. These materials present two magnetic structures and are very poor semiconductors; many authors approximate them as insulators. In any case, their conductivity is very small. However, metals have the largest conductivities and show particular interactions with microwaves. The conductivity of a metal, or conducting material, presents millions of "free" electrons to the actions of the electric, $J = \sigma E$, and magnetic fields, $q\mathbf{v}\times\mathbf{B}$, of the microwaves, and the Lorentz force makes them jiggle rapidly in the resistive medium they are, generating Joule losses and eventually heat. The higher the frequency of the microwaves, the conductivity can become frequency-dependent. In general, the responses $\epsilon(\omega)$, $\mu(\omega)$, and $\sigma(\omega)$ are heavily dependent on the frequency of the microwaves. This is the dispersion characteristic of matter. As an example, clouds, water vapor, hot air, and sea water have quite different electrodynamic properties; that is why weather radar of longest wavelength (L-band, $\lambda \approx 1$ m, $\nu \approx 1.24$ GHz) is used routinely to penetrate clouds, and weather radar of shorter wavelengths (C-band, $\lambda \approx 0.25$ m, $\nu \approx 3.2$ GHz) is used to detect forests, and even X-band radar is used to measure altitude with respect to sea level due to its highest reflectivity on the surface of the ocean. The $\epsilon$, the $\mu$, and the $\sigma$ of a material take away, absorb, energy any time they are microwave bathed, much the same way as when we heat coffee or meals in the microwave oven.

We describe here, in a nutshell form, the fundamental physics involved in the interactions of microwaves with matter with emphasis on magnetic matter. Today science and technology are able to produce, control, direct, and amplify microwaves in the laboratory and in devices and high-technology instruments and equipments [9, 11, 19, 21]. The electromagnetism of microwave devices and instruments is an integral part of today's technological world. And the electrodynamics of microwaves is governed by Maxwell's equations as applied to $\epsilon$, $\mu$, and $\sigma$ materials. We present these equations and the partial differential wave equations that are obtained from them with the expressions of their solutions.

## 3. Maxwell's equations are the universe of electromagnetic (electrodynamic) phenomena

Richard Feynman (from Feynman's Lectures, Vol. 2, Chap. 21): "So here is the center of the universe of electromagnetism, the complete theory of electricity and magnetism, and of light; a complete description of the fields produced by any moving charges; and more. It is all here. Here is the structure built by Maxwell, complete in all its power and beauty. It is probably one of the greatest accomplishments of physics. To remind you of its importance, we will put it all together in a nice frame:"

$$\nabla \cdot \mathbf{E} = \frac{1}{\epsilon_0}(\rho_f - \nabla \cdot \mathbf{P}) \tag{1}$$

$$\nabla \cdot \mathbf{B} = 0 \tag{2}$$

$$\nabla \times \mathbf{E} + \frac{\partial \mathbf{B}}{\partial \mathbf{t}} = \mathbf{0} \tag{3}$$

$$\nabla \times \mathbf{B} - \frac{1}{c^2}\frac{\partial \mathbf{E}}{\partial t} = \mu_0\left(\mathbf{J}_f + \frac{\partial \mathbf{P}}{\partial t} + \nabla \times \mathbf{M}\right) \tag{4}$$

He states such definite statements once he obtains the general solutions for the electric potential, V, and the vector potential, **A**, for charges in motion. We have changed Feynman's notation $\varphi$ for V to denote electric potential, and we use MKS units. In these equations $\rho$'s are charge densities, **J**'s are free currents, **P** is the polarization vector, **M** the magnetization vector, and **E** and **B** the electric and magnetic fields. We already defined $\epsilon$, $\mu$, and $\sigma$ above.

We are not concerned here in the detail deduction of Maxwell's equations, nor the deduction of electromagnetic waves from them. Reflection and refraction and absorption of microwaves in conducting magnetic and dielectric materials are presented in graphical form along with the principal equations that govern them. The deduction of such results is too long, and they are developed in a number of excellent, well-established textbooks [27–33]. The deduction of electromagnetic fields and waves from Maxwell's equations does not impose any restriction on the wavelength-frequency, as shown in **Figure 1**. The result is quite general, but we are interested and focused on microwaves interacting with $\epsilon$, $\mu$, and $\sigma$ matter. But real materials, the air, ocean, hurricanes, semiconductors, ferrites, conductors, submarines, cars, planes, ferrous materials, steel, and magnets, which are of technological relevance, are dispersive, anisotropic, and absorb microwave energy [19, 28, 31]; then the adequate $\epsilon$ and $\mu$ become complex (dispersion and dissipation effects taken into account) and tensorial (anisotropy effects taken into account) [28, 31, 32]:

$$\boldsymbol{\epsilon_{ij}}=\begin{bmatrix}\epsilon_{11} & \epsilon_{12} & \epsilon_{13}\\ \epsilon_{21} & \epsilon_{22} & \epsilon_{23}\\ \epsilon_{31} & \epsilon_{32} & \epsilon_{33}\end{bmatrix}=\begin{bmatrix}\epsilon_{x} & 0 & j\epsilon_{13}\\ 0 & \epsilon_{y} & 0\\ -j\epsilon_{31} & 0 & \epsilon_{z}\end{bmatrix} \tag{5}$$

$$\boldsymbol{\mu_{ij}}=\begin{bmatrix}\mu_{11} & \mu_{12} & \mu_{13}\\ \mu_{21} & \mu_{22} & \mu_{23}\\ \mu_{31} & \mu_{32} & \mu_{33}\end{bmatrix}=\begin{bmatrix}\mu & 0 & jk\\ 0 & 1 & 0\\ -jk & 0 & \mu\end{bmatrix} \tag{6}$$

All the elements of both material electrodynamic tensors are in general dependent on frequency, $\boldsymbol{\epsilon}_{ij}(\omega)$ and $\boldsymbol{\mu}_{ij}(\omega)$. In what follows we will continue to use just the symbols $\epsilon$ and $\mu$. However, we understand their complexity. Using all four field vectors **E**, **D**, **B** and **H**,

$$\nabla\cdot\mathbf{D}=\rho_f \tag{7}$$

$$\nabla\cdot\mathbf{B}=0 \tag{8}$$

$$\nabla\times\mathbf{E}+\frac{\partial\mathbf{B}}{\partial\mathbf{t}}=\mathbf{0} \tag{9}$$

$$\nabla\times\mathbf{H}-\frac{\partial\mathbf{D}}{\partial\mathbf{t}}=\mathbf{J_f} \tag{10}$$

Remember that **D** is $\epsilon_0\mathbf{E}+\mathbf{P}$ and $\mathbf{P}=\epsilon_0\chi_e\mathbf{E}$, while **B** is $\mu_0(\mathbf{H}+\mathbf{M})$, and $\mathbf{M}=\chi_m\mathbf{H}$, and $\mu=\mu_0\mu_r=\mu_0(1+\chi_m)$. Notice that the material's electrodynamics response is within **D** and **B**. In fact, the fields **D** and **B** are valued inside the material bodies. We are interested in microwave fields that in free space and in $\varepsilon$, $\mu$, and $\sigma$ matter, the field components **E** and **H** are sinusoidal functions of time and orthogonal to each other. In such case we can replace the time derivative by factors of $j\omega$. Setting also $\mathbf{D}=\epsilon\mathbf{E}$ and $\mathbf{B}=\mu\mathbf{H}$,

$$\nabla\cdot\epsilon\mathbf{E}=\rho_f \tag{11}$$

$$\nabla\cdot\mu\mathbf{H}=0 \tag{12}$$

$$\nabla\times\mathbf{E}+j\omega\mu\mathbf{H}=\mathbf{0} \tag{13}$$

$$\nabla \times \mathbf{H} - j\omega\epsilon\mathbf{E} = \mathbf{J}_f \tag{14}$$

From Maxwell' equations inhomogeneous wave equations are obtained for the fields **E** and **H**. The procedure is taking the rot ($\nabla\times$) of the two rot Maxwell equations, rearranging terms and using the divergence equations; we find for a medium $\epsilon$, $\mu$, and $\sigma$ nearby ($\rho_f, \mathbf{J}_f$) microwave source:

$$\nabla^2\mathbf{E} - \epsilon\mu\frac{\partial^2\mathbf{E}}{\partial t^2} = \nabla\left(\frac{\rho_f}{\epsilon}\right) + \mu\frac{\partial\mathbf{J}_f}{\partial t} \tag{15}$$

$$\nabla^2\mathbf{B} - \epsilon\mu\frac{\partial^2\mathbf{B}}{\partial t^2} = \mu\nabla\times\mathbf{J}_f \tag{16}$$

Far from the microwave sources (centimeters, meters, kilometers, and, even, light years), we can safely drop the "source" terms, and homogeneous partial differential equations are left:

$$\nabla^2\mathbf{E} - \epsilon\mu\frac{\partial^2\mathbf{E}}{\partial t^2} - \mu\sigma\frac{\partial\mathbf{E}}{\partial t} = 0 \tag{17}$$

$$\nabla^2\mathbf{H} - \epsilon\mu\frac{\partial^2\mathbf{H}}{\partial t^2} - \mu\sigma\frac{\partial\mathbf{H}}{\partial t} = 0 \tag{18}$$

It should be stressed that these identical wave equations describe attenuated waves. This comes naturally from the third term that includes magnetic permeability, electric conductivity, and the first time derivative of each field. The solutions of these equations are readily obtained by separation of variables [27]. The detailed deduction of these mathematical expressions is lengthy and elaborated; after such work, we find the fields to be transverse and plane-polarized for which the x-direction for **E** is chosen; hence **H** ($(j\omega\mu)\mathbf{H} = \nabla \times \mathbf{E}$) goes along the y-direction and propagates along the z-direction with a flux of energy density $\mathbf{S} = \mathbf{E} \times \mathbf{H}$ [27] and an impedance Z = E/H. The solutions are summarized in **Table 1**.

Here we show in the form of a table (deductions are given in [27, 28, 31]) the solutions for E and H for propagation in free space and in a dielectric material and in a conducting material. In this concise form, we highlight the most relevant features and parameters of the solutions. When we put back the solutions found for a $\sigma$, $\mu$, and $\epsilon$ material into the wave equations, we obtain the dispersion relation: $(-k^2 + \omega^2\epsilon\mu - j\omega\sigma\mu)\mathbf{E} = 0$, which is shown in the third column for each

| Propagation | Equation | Dispersion Relation | Impedance $Z$[ohms] | Some Parameters |
|---|---|---|---|---|
| Vacuum | $E = E_m \exp j(\omega t - kz)$<br>$H = H_m \exp j(\omega t - kz)$ | $(-k^2 + \omega^2\epsilon_0\mu_0)E = 0$ | $Z_0 = \frac{E}{H} = \frac{k}{\omega\epsilon_0} = \frac{\omega\mu_0}{k} = \frac{1}{\epsilon_0 c} = \mu_0 c = \left(\frac{\mu_0}{\epsilon_0}\right)^{1/2}$ | $k \to$ real<br>$Z \to$ real<br>Skin Depth $\delta = 0$<br>No losses |
| Nonconductors | $E = E_m \exp j(\omega t - kz)$<br>$H = H_m \exp j(\omega t - kz)$ | $(-k^2 + \omega^2\epsilon\mu)E = 0$ | $Z = \frac{E}{H} = \left(\frac{\mu}{\epsilon}\right)^{1/2} = 377\left(\frac{\mu_r}{\epsilon_r}\right)^{1/2}$ | $k \to$ real<br>$Z \to$ real<br>Skin Depth $\delta = 0$<br>No losses |
| Magnetic Conductors, possibly dielectric | $E = E_m \exp(-\alpha z) \exp j(\omega t - \beta z)$<br>$H = H_m \exp(-\alpha z) \exp j(\omega t - \beta z - \theta)$ | $(-k^2 + \omega^2\epsilon\mu - j\omega\sigma\mu)E = 0$ | $Z = \frac{E}{H} = \frac{k}{\omega\epsilon - j\sigma} = \frac{\omega\mu}{k}$<br>$= \left(\frac{\mu}{\epsilon}\right)^{1/2} \frac{\exp j \arctan(\alpha/\beta)}{(1 + D^2)^{1/4}}$<br>$\approx 377\left(\frac{\mu_r}{\epsilon_r}\right)^{1/2} \frac{\exp j \arctan(\alpha/\beta)}{(1 + D^2)^{1/4}}$ | $k = \beta - j\alpha \to$ necessary complex<br>$\alpha$ gives absorption; $\alpha \neq \beta$<br>$z \to$ necessary complex<br>$\delta = \frac{1}{\alpha} = \left(\frac{2}{\sigma\mu\omega}\right)^{1/2}$ |
| Very Good Conductors | $E = E_m \exp\left[j\left(\omega t - \frac{z}{\delta}\right) - \frac{z}{\delta}\right]$<br>$H = \left(\frac{\sigma}{\omega\mu}\right)^{1/2} E_m \exp\left[j\left(\omega t - \frac{z}{\delta} - \frac{\pi}{4}\right) - \frac{z}{\delta}\right]$ | $(-k^2 + \omega^2\epsilon_0\mu_0 - j\omega\sigma\mu_0)E = 0$ | $Z = 377\left(\frac{\mu_0}{\epsilon_0}\right)^{1/2} \frac{1}{(1 + D^2)^{1/4}}$ | $k = \beta - j\alpha \to$ necessary complex<br>$\alpha$ gives absorption; $\alpha = \beta$<br>$z \to$ becomes real<br>$\delta = \frac{1}{\alpha} = \left(\frac{2}{\sigma\mu_0\omega}\right)^{1/2}$ |

**Table 1.**
*Wave solutions in different media (no bounds considered).*

electrodynamic medium. A summary of **E** and **H** solutions, along with the dispersion relations, impedance, attenuation, and skin depth, is presented in **Table 1** for different materials characterized by particular sets of electrodynamic properties ($\sigma$, $\mu$, $\epsilon$).

In the ideal case of propagation of microwaves in air, the phenomenon of attenuation (absorption of microwave energy, $E = h\nu$, in the form of photons) does not take place. In reality, quite a substantial attenuation of microwaves takes place when they propagate in air (molecules that constitute the air readily absorb microwaves. Such energy excites efficiently rotation and vibration degrees of freedom). The Wi-Fi technology [5, 6, 16] knows very well this fact, that is why there exist in the market potentiators of Wi-Fi signal to cover adequately a room of several meters square. The problem of attenuated signal becomes worse if panels, wall, wood, and metals come in the way of the propagation. The microwave engineer, designer, and scientist know that the main factor which causes attenuation in microwave propagation is water, $H_2O$. War technology for decades has faced the problem of the no radar signal penetration in the oceans to a depth enough to detect submarines. So, in order to propagate microwaves in the atmosphere or in the ocean, for at least 1 m, or longer distances, it is mandatory to guide such electromagnetic GHz waves or increase considerably their power content, Poynting vector, $\mathbf{S} = \mathbf{E} \times \mathbf{H}$ being proportional to its power content per unit area.

### 3.1 Microwaves in spaces with obstacles: propagation, reflection, refraction, and absorption

In the modern microwave Wi-Fi, military, radar communication systems, the microwaves have to travel through "free space" to reach detectors and receivers, but it is pretty common in urban areas that the propagation encounters obstacles, such as buildings, windows, walls, bridges, metallic structures, and fog. At the surface of those obstacles, reflection, absorption, refraction, and dispersion of the microwaves occur. When microwaves hit obstacles, energy is irremediably lost at these obstacles. This fact could be used to detect that precisely an obstacle is present and fine measurements of the reflected or refracted microwave energy could give valuable information on the electromagnetic character ($\epsilon$, $\mu$, $\sigma$) of such obstacle. This is how radar in air and other atmospheric microwave instruments detect objects (the obstacles themselves). But many other times, obstacles, like walls, wood floors, and concrete structures, attenuate, disperse, and deviate the microwaves, and reception at the desired site is deficient or completely lost. In such cases midway potentiators are used to increase the signal power to compensate the losses at the obstacles. These examples are meant to show how important are the reflection, absorption, refraction, and dispersion of microwaves in real-world applications. The laws that govern these phenomena are Snell law and Fresnel laws of reflection and refraction. They are obtained as a consequence of the boundary conditions applied to the interface, any interface between two media as shown in **Figure 4**. The reflected (transmitted) waves carry an energy R(T) per unit of incident energy. Appropriate detectors (common radiometers, photocells, diodes) measure with ease R and T, also depicted in **Figure 4**. These radar and Wi-Fi microwaves hit all the time buildings, hot air, street floors, and so forth, and, at all times, they are reflected, absorbed, and refracted [16–20] as illustrated in **Figure 4**.

### 3.2 Boundary conditions that microwaves fulfill at the interface between any two media

The general boundary conditions the electromagnetic fields, **E**, **D**, **H** and **B**, must satisfy at the interface of any two ($\sigma$, $\mu$, $\varepsilon$) media which are easily obtained

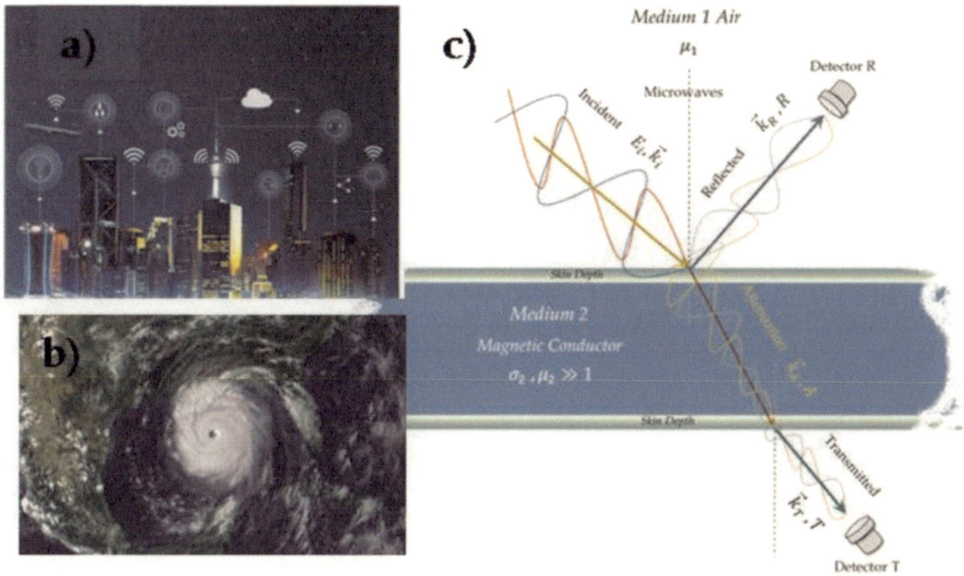


**Figure 4.**
*Microwaves bounce back from virtually any material. (a) The Wi-Fi signal being reflected and refracted from anything can be in an urban city area. Wi-Fi and radar signals are commonly present in the air in any city. These microwaves hit buildings, windows, fog, hot air, street floors, and so forth; at all times they are reflected (R), absorbed (A), and refracted (T) [16–20], (b) satellite weather radar detecting and following 2005 Katrina hurricane trajectory just about to land [21], and (c) abstraction of any process in which microwaves hit a body of matter.*

from applying Gaus law to a small cylinder placed through the two media and from applying Ampere's circuital law to a closed (rectangular) trajectory, again, crossing through both media. The results are [27–29] $E_{1t} = E_{2t}$; $H_{1t} = H_{2t}$, i.e., the tangential components of **E** and **H** must be continuous through the interface. And $D_{1n} = D_{2n}$; $B_{1n} = B_{2n}$, i.e., the normal components of **D** must be continuous, provided the interface does not carry any charge and the normal components of **B** must be continuous. All this is represented in **Figure 5**.

Since medium 1 sustains the incident and reflected fields, $\mathbf{E}_1 = \mathbf{E}_i + \mathbf{E}_r$; $\mathbf{D}_1 = \mathbf{D}_i + \mathbf{D}_r$; $\mathbf{H}_1 = \mathbf{H}_i + \mathbf{H}_r$; and $\mathbf{B}_1 = \mathbf{B}_i + \mathbf{B}_r$. In medium 1 a sum of vector fields must be formed. In medium 2 there are only the transmitted fields $\mathbf{E}_t$, $\mathbf{D}_t$, $\mathbf{H}_t$, and $\mathbf{B}_t$.

When boundaries are present on purpose or as an inevitable obstacle, the incident microwaves reflect and refract, and the formal treatment will conduct us to Snell law and the Fresnel equations for incidence with **H** parallel to the plane of incidence as shown in **Figures 6** and **7** to the Fresnel equations with **H** perpendicular to the plane of incidence as shown in **Figure 8**.

In **Figure 6a**, medium 1 is the space where we want to propagate some microwave signal and medium 2 being a window, a wall, water, wood, concrete, metallic sheets, a building in front of us, etc. A very common situation is when sunlight shines and goes through a glass window; in this case most of the light goes through and illuminates the room. And yet, part of it is reflected, and part is absorbed (the energy of which heats up the glass window itself). A second very common situation generated by the modern wireless communication technology is the Wi-Fi signal (it plays exactly the same role as the sunshine in the above illustration). The typical Wi-Fi signal is around 2.3 GHz, and the newest cable boxes are delivering to our home rooms ≈5.2GHz with about 1 W power.

When the obstacle (medium 2) is made of materials that are also ferromagnetic or ferrimagnetic (ferrites), the absorption will increase considerably because the magnetic moments of the material absorb energy from the Wi-Fi waves.

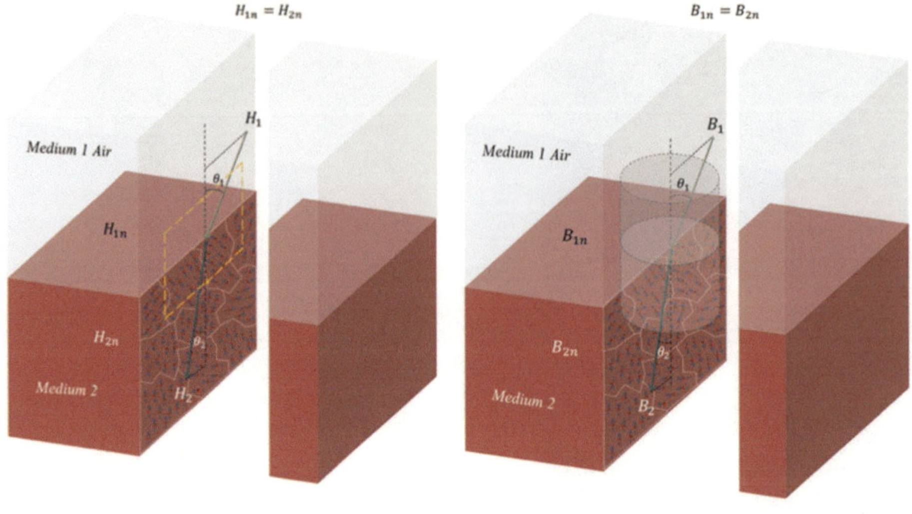


**Figure 5.**
*Boundary conditions for the magnetic field* **H** *and for the magnetic induction. The condition* $B_{1n} = B_{2n}$ *is obtained from* $\nabla \cdot \mathbf{B} = 0$, *and* $H_{1t} = H_{2t}$ *is obtained from Ampere's circuital law. Similar vector diagrams for the* **D** *and* **E** *vectors are considered, and the conditions,* $D_{1n} = D_{2n}$ *and* $H_{1t} = H_{2t}$, *are obtained.*

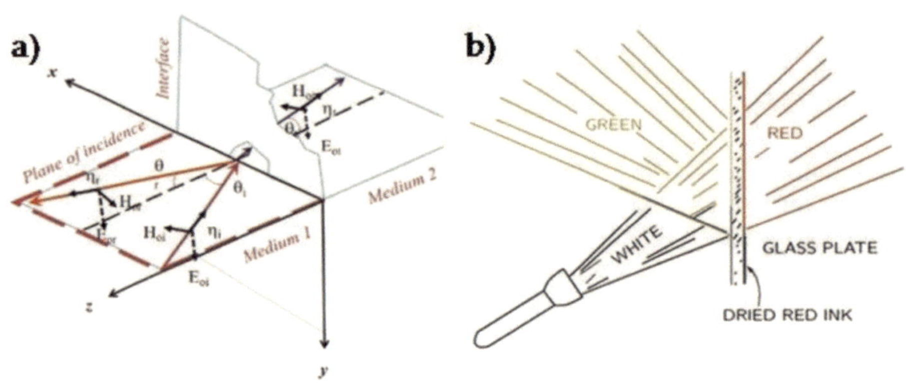


**Figure 6.**
*Incident, reflected, and transmitted microwaves at any kind of interface. (a)* **H** *parallel to the plane of incidence.* $\mathbf{n}_i$, $\mathbf{n}_r$, *and* $\mathbf{n}_t$ *are unitary vectors normal to the respective front-plane waves and pointing in the propagation direction,* $\mathbf{k} = k\mathbf{n}$. $\theta_i$, $\theta_r$, and $\theta_t$ *are, respectively, the incident, the reflected, and the transmitted direction of the waves with respect to the dotted line axis. (b) Feynman's Lintern. A white light beam incident on a glass that has been coated with red paint refracts red light and reflects the green component; hence* $\omega_t \neq \omega_r$.

Conductors (materials with free electrons) will absorb Wi-Fi wave energy by virtue of the term JE which becomes Joule heating. Abstracting the cases mentioned above, we consider, as the first case, the incidence of a plane electromagnetic wave on the interface of two media as shown in **Figure 7**; the incident electric field is of the form

$$\mathbf{E_i} = \mathrm{E_{0i}} \exp j\omega_i \left( t - \frac{\mathbf{n_i} \cdot \mathbf{r}}{u_1} \right) \mathbf{n}_i \tag{19}$$

The reflected and the refracted waves are $\mathbf{E_r} = \mathbf{E_{0r}} \exp j\omega_r \left( t - \frac{\mathbf{n_r} \cdot \mathbf{r}}{u_1} \right)$ and $\mathbf{E_t} = \mathbf{E_{0t}} \exp j\omega_t \left( t - \frac{\mathbf{n_t} \cdot \mathbf{r}}{u_2} \right)$, where $u_1$and $u_2$ are the velocities of the waves in mediums 1 and 2, respectively. The $E_{0i}$, $E_{0r}$, and $E_{0r}$ are the amplitudes of the respective

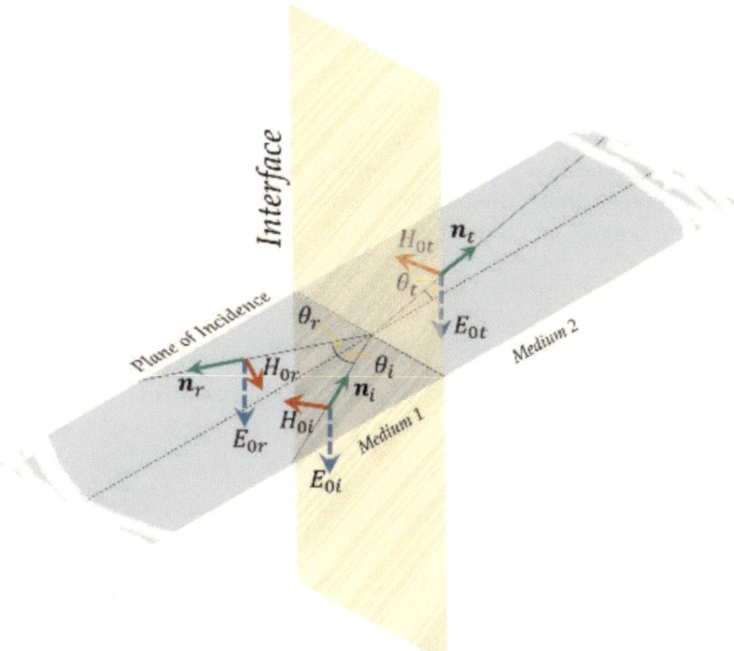


**Figure 7.**
*The incident wave with magnetic vector in the incidence plane. Snell laws demand that the reflected and transmitted magnetic vectors are also in the same plane. The arrows indicate the directions of the vectors involved. The Poynting vectors $\mathbf{S}_{i,r,t} = \mathbf{E}_{i,r,t} \times \mathbf{H}_{i,r,t}$ point always in the direction of propagation of that particular wave, and its magnitude indicates how much electromagnetic energy is flowing in that direction.*

waves. For linear, isotropic, homogeneous media (LIH), and nondispersive, it is readily shown that the three vector fields $\mathbf{E}_i$, $\mathbf{E}_r$, and $\mathbf{E}_t$ are identical functions of time [27–29, 32–34], $\omega_i = \omega_r = \omega_t$. For dispersive media, at least [27–29, 32–34] two frequencies are different $\omega_i \neq \omega_t \neq \omega_r$.

As the simple example illustrated in **Figure 6b**, given by Feynman [29], shows "A large imaginary part of the index of refraction (or equivalently a large imaginary component of k, k") means a strong absorption. So there is a general rule that if any material gets to be a *very good absorber at any frequency (let's say red)*, the waves are strongly reflected at the surface, and very little gets inside to be absorbed. You can see this effect with strong dyes. Pure crystals of the strongest dyes have a metallic shine. Red ink absorbs out the greens of transmitted light, so if the ink is very concentrated, it will exhibit a strong surface reflection for the frequencies of green light (see **Figure 6b**).

But in our case the three fields must be identical functions of space and time at any point on the interface [27–29, 32, 33]. From such requirement, it follows that $\theta_i = \theta_r$ the angle of incidence is equal to the angle of reflection. This is the law of reflection. It is also obtained that $[\sin\theta_t/\sin\theta_i] = [n_1/n_2]$, with $n = k/\lambda$. This equation is the well-known Snell law. With these results at hand, we now write: $\mathbf{E}_i = \mathbf{E}_{0i} \exp j[\omega t - k_1(x\sin\theta_i - z\cos\theta_i)]$, $\mathbf{E}_r = \mathbf{E}_{0r} \exp j[\omega t - k_1(x\sin\theta_i - z\cos\theta_i)]$, $\mathbf{E}_t = \mathbf{E}_{0t} \exp j[\omega t - k_2(x\sin\theta_t - z\cos\theta_t)]$; the respective magnetic fields are $H_p = (k/\omega\mu)E_p$, with $p = i, r, t$.

Now we proceed to determine the quantities $E_{0i}$, $E_{0r}$, $E_{0t}$ that assure the continuity of the tangential components of **E** and **H** at the interface:

$\mathbf{E} = E_0 \exp j(\omega t - \mathbf{k}_\eta \cdot \mathbf{r})\mathbf{n}_\eta$ and $\mathbf{H} = (k_\eta/\omega\mu_{1,2})E_0 \exp j(\omega t - \mathbf{k}_\eta \cdot \mathbf{r})\mathbf{n}_\eta$, with $\mathbf{k}_\eta = \mathbf{k}_i$, $\mathbf{k}_r$, $\mathbf{k}_t$ with $\eta$ = i, r, t the propagation vectors and $\mu_{1,2}$ the magnetic permeability of medium 1 or medium 2.

We have four unknowns: ($E_{0r}$, $H_{0r}$) and ($E_{0t}$, $H_{0t}$). And we have four equations, which are the four boundary conditions. $E_{0i}$ and $H_{0i}$ are taken as known since they are the primary waves that we send-propagate from a "controllable" source. We would find the unknowns normalized by $E_{0i}$ and $H_{0i}$. Of course the problem can be inverted, and we could start knowing the transmitted waves and would like to determine the initial fields, $E_i$, $H_i$, that come from an unknown (potentially fundamental) source. Present cosmological problems are exactly of this type [3, 35, 36]. Continuity of the tangential components of **E** requires

$$E_{0i} + E_{0r} = E_{0t} \tag{20}$$

At any point and at any time at the interface. Likewise, continuity of the tangential component of the magnetic field requires $H_{0i}\cos\theta_i - H_{0r}\cos\theta_i = H_{0t}\cos\theta_t$ which becomes $\frac{k_1}{\omega\mu_1}(E_{0i} - E_{0r})\cos\theta_i = \frac{k_2}{\omega\mu_2}E_{0t}\cos\theta_t$. But $k = n/\lambda$, so

$$\frac{n_1}{\mu_1}(E_{0i} - E_{0r})\cos\theta_i = \frac{n_2}{\mu_2}E_{0t}\cos\theta_t \tag{21}$$

Algebraic Eqs. (20) and (21) are readily solved for unknowns $E_{0r}$ and $E_{0t}$:

$$\frac{n_1}{\mu_1}E_{0i}\cos\theta_i - \frac{n_1}{\mu_1}E_{0r}\cos\theta_i = \frac{n_2}{\mu_2}E_{0t}\cos\theta_t \tag{22}$$

Using (20) $E_{0i} + E_{0r} = E_{0t}$ and after some algebra, we obtain [27]: the amplitude of the fields $E_{0r}/E_{0i}$ and $E_{0t}/E_{0i}$:

$$\left(\frac{E_{0r}}{E_{0i}}\right)_N = \frac{\frac{n_1}{\mu_{r1}}\cos\theta_i - \frac{n_2}{\mu_{r2}}\cos\theta_t}{\frac{n_1}{\mu_{r1}}\cos\theta_i + \frac{n_2}{\mu_{r2}}\cos\theta_t} \tag{23}$$

$$\left(\frac{E_{0t}}{E_{0i}}\right)_N = \frac{2\frac{n_1}{\mu_{r1}}\cos\theta_i}{\frac{n_1}{\mu_{r1}}\cos\theta_i + \frac{n_2}{\mu_{r2}}\cos\theta_t} \tag{24}$$

where N indicates that E is normal to the incidence plane and/or H is in the plane of incidence.

In the case the three H vectors are perpendicular to the plane of incidence, **Figure 8**, we have: $H_{0i} - H_{0r} = H_{0t}$. In terms of **E** fields

$$\frac{n_1}{\mu_1}(E_{0i} - E_{0r}) = \frac{n_2}{\mu_2}E_{0t} \tag{25}$$

$$(E_{0i} + E_{0r})\cos\theta_i = E_{0t}\cos\theta_t \tag{26}$$

After some algebra and rearranging terms, we obtain $E_{0t}/E_{0i}$. Hence:

$$\left(\frac{E_{0r}}{E_{0i}}\right)_P = \frac{-\frac{n_2}{\mu_{r2}}\cos\theta_i + \frac{n_1}{\mu_{r1}}\cos\theta_t}{\frac{n_1}{\mu_{r1}}\cos\theta_t + \frac{n_2}{\mu_{r2}}\cos\theta_i} \tag{27}$$

$$\left(\frac{E_{0t}}{E_{0i}}\right)_P = \frac{2\frac{n_1}{\mu_{r1}}\cos\theta_i}{\frac{n_1}{\mu_{r1}}\cos\theta_t + \frac{n_2}{\mu_{r2}}\cos\theta_i} \tag{28}$$

Here P denotes that the $\mathbf{H}_i$, $\mathbf{H}_r$, $\mathbf{H}_t$ are all parallel to the interface. In the case of normal incidence with all the $\mathbf{H}$ vectors parallel to the interface, $\theta_i = \theta_t = 0$ (**Figure 8**), and $\cos\theta_i = \cos\theta_t = 1$ in the above equations.

### 3.3 The coefficients of reflection (R) and transmission (T) and conservation of energy (R + T = 1)

$$R = \left[\frac{S_{r\ av}\cdot \mathbf{n}}{S_{i\ av}\cdot \mathbf{n}}\right] = \frac{E^2_{0\ r}}{E^2_{0\ t}} \tag{29}$$

$$T = \left[\frac{\mathbf{S}_{t\ av}\cdot \mathbf{n}}{\mathbf{S}_{i\ av}\cdot \mathbf{n}}\right] = \left(\frac{\epsilon_{r\ 2}}{\epsilon_{r\ 1}}\right)^{1/2}\frac{E^2_{0\ t}\cos\theta_t}{E^2_{0\ t}\cos\theta_i} = \frac{n_2E^2_{0\ t}\cos\theta_t}{n_1E^2_{0\ t}\cos\theta_i} \tag{30}$$

$$R_N = \left\{\frac{\left(\frac{n_1}{n_2}\right)\cos\theta_i - \cos\theta_t}{\left(\frac{n_1}{n_2}\right)\cos\theta_i + \cos\theta_t}\right\}^2 \tag{31}$$

$$T_N = \frac{4\left(\frac{n_1}{n_2}\right)\cos\theta_i\cos\theta_t}{\left\{\left(\frac{n_1}{n_2}\right)\cos\theta_i + \cos\theta_t\right\}^2} \tag{32}$$

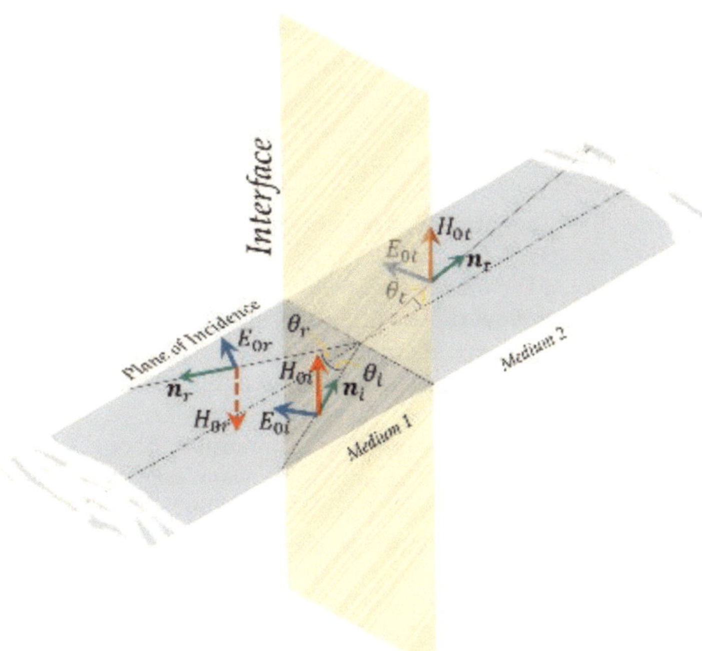


**Figure 8.**
*Both media are dielectric and magnetic, and the magnetic field component $\mathbf{H}_i$ is parallel to the interface, and so the reflected and the transmitted magnetic components, $\mathbf{H}_r$, $\mathbf{H}_t$, are also parallel to the interface.*

$$R_P = \left\{ \frac{-\cos\theta_i + \left(\frac{n_1}{n_2}\right)\cos\theta_t}{\cos\theta_i + \left(\frac{n_1}{n_2}\right)\cos\theta_t} \right\}^2 \quad (33)$$

$$T_p = \frac{4\left(\frac{n_1}{n_2}\right)\cos\theta_i\cos\theta_t}{\left\{\cos\theta_i + \left(\frac{n_1}{n_2}\right)\cos\theta_t\right\}^2} \quad (34)$$

And obviously $R_N + T_N = 1$, and $R_p + T_p = 1$. With these equations we calculate the energy content in the reflected wave and in the transmitted wave in terms of the incident wave for both cases: the so-called parallel incidence and the normal incidence. For normal incidence as in geophysics altimeter radars, $\theta_i = \theta_r = \theta_t = 0^0$, and the R and T coefficients become $R = \{(n_1 - n_2)/(n_1 + n_2)\}^2$ and $T = (4n_1)/\{(n_1 + n_2)\}^2$.

The whole subject of reflection, refraction, and absorption of microwaves at the interface of air and a conducting ($\sigma$), magnetic ($\mu$), and dielectric ($\epsilon$) medium is summarized qualitatively in **Figure 4**. The most relevant mathematical expressions are given above. Depending on the application at hand, reflection and transmission count with sensitive methods for their measurement, and in consequence quantitative determinations of absorption of microwave power is readily available.

## 4. Waveguides

When research in this area does not involve necessarily open spaces, and transmission losses should be avoided, what can we do to reduce losses, dispersion, and uncontrolled reflections of microwaves while propagating? Or what can be done to control and measure such dispersions, absorptions, and reflections? The answer is whenever possible, guide the microwaves (**Figure 9**).

It is well known that very good conductors (metallic) reflect electromagnetic waves with a minimum of losses (these losses are due to Joule effect on the free electrons that are within the skin depth only) (see **Table 1**). And this skin depth is very small (microns, fractions of microns) for good conductors. So, multiple reflections on hollow metallic pipes are preferred choice to deliver microwaves from here to there (**Figures 9** and **10**).

Since the beginnings of the microwave technology, previous to world war II, it is well known that metallic hollow pipes with internal, mirror-polished walls can sustain propagation of some particular electromagnetic (EM) modes, TM (transverse magnetic), and TE (transverse electric) and cannot sustain other EM modes [27–29, 37, 38].

A universal condition is that one significant dimension, $\xi$, of the hollow pipe be exactly a multiple of an integer number of half the wavelength of the microwave to be transmitted through it. Hence $\xi = n\lambda/2$ determines the "size" of the cross section of a rectangular or a cylindrical waveguide. $\lambda_g = c/\nu_{nm}$ is the wavelength inside the waveguide. All the theory is consequence of the solutions to Maxwell's equations, under boundary conditions at the walls of the mirror-polished metallic surfaces of the microwaveguide. The particular deductions of the mathematical expressions of the valid **E** and **H** fields inside the waveguides are involved and lengthy. We give in **Table 2** some rectangular waveguides with their band, frequency operation, cutoff frequencies, and internal dimensions (**Figure 10**). A typical rectangular Q-band waveguide connected to a Q-cylindrical resonant cavity is shown in **Figure 10b**.

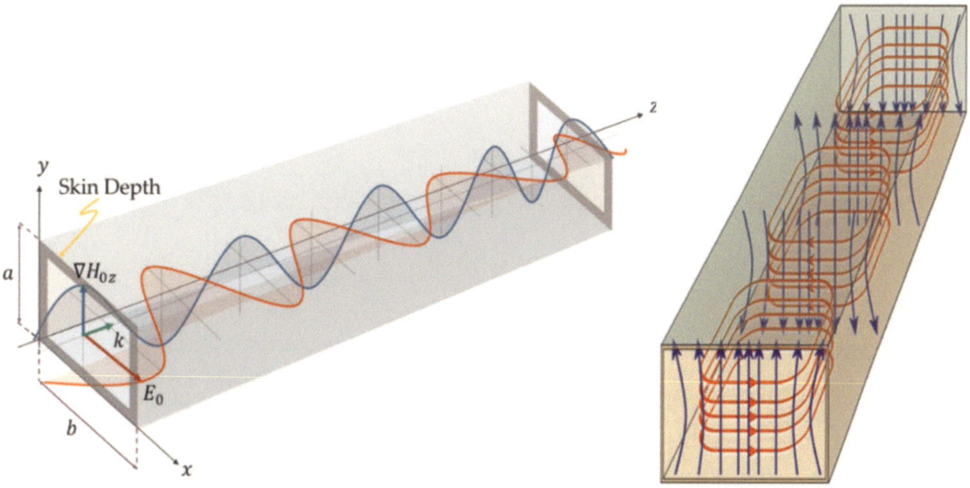


**Figure 9.**
*Rectangular waveguides made of very good conducting metals (copper, silver, gold, and brass) sustaining different patterns of **E** and **H** modes propagating along the guides. If the guide is perfectly conducting the electric field intensity, **E** is zero in the conductor, and **E** is either normal or zero at the surface. For a TE wave, it is shown that $\nabla H_{0z}$ is tangent to the wall.*

| Standard sizes of rectangular waveguide | | | | | |
|---|---|---|---|---|---|
| **Waveguide name** | **Frequency band name** | **Recommended frequency band of operation (GHz)** | **Cutoff frequency of lowest-order mode (GHz)** | **Cutoff frequency of next mode (GHz)** | **Inner dimensions of waveguide opening (mm)** |
| **EIA** | | | | | |
| WR650 | L-band (part) | 1.15–1.72 | 0.908 | 1.816 | 165.1 × 82.55 |
| WR340 | S-band (part) | 2.20–3.30 | 1.736 | 3.471 | 86.36 × 43.18 |
| WR229 | C-band (part) | 3.30–4.90 | 2.577 | 5.154 | 58.17 × 29.08 |
| WR90 | X-band | 8.20–12.40 | 6.557 | 13.114 | 22.9 × 10.2 |
| WR42 | K-band | 18.00–26.50 | 14.051 | 28.102 | 10.7 × 4.32 |
| WR22 | Q-band | 33.00–50.00 | 26.346 | 52.692 | 5.86 × 2.84 |
| WR19 | U-band | 40.00–60.00 | 31.391 | 62.782 | 4.78 × 2.39 |
| WR15 | V-band | 50.00–75.00 | 39.875 | 79.75 | 3.76 × 1.88 |
| WR12 | E-band | 60.00–90.00 | 48.373 | 96.746 | 3.10 × 1.55 |
| WR10 | W-band | 75.00–110.00 | 59.015 | 118.03 | 2.54 × 1.27 |
| WR8 | F-band | 90.00–140.00 | 73.768 | 147.536 | 2.03 × 1.02 |
| WR6, WR7, WR6, 5 | D-band | 110.00–170.00 | 90.791 | 181.583 | 1.65 × 0.826 |

**Table 2.**
*Some of the most common rectangular waveguides and their frequency ranges and frequency cutoffs and inner dimensions (mm). The waveguide name WR stands for waveguide rectangular, and the number is the inner dimension width of the waveguide in hundredths of an inch (0.01 inch = 0.254 mm). The different microwave bands are given and can be correlated with the bands shown in **Figure 1** (Taken from Wikipedia).*

Once we know the electromagnetic patterns that can be formed and sustained in hollow metallic pipes, as the one we show in **Figure 11**, how can we use them to probe material's properties? The fundamental idea of how to measure

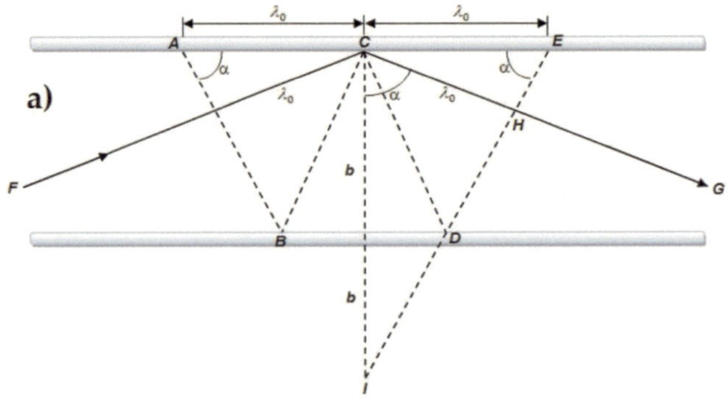


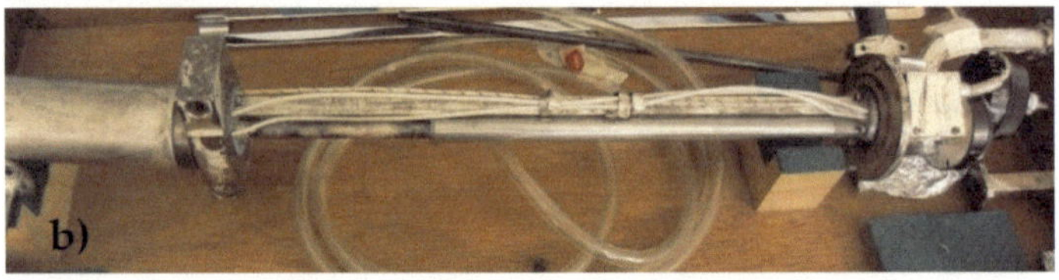


**Figure 10.**
*A plane electromagnetic wave propagating in a rectangular hollow waveguide. (a) The lines AB and CD are parallel to wave fronts for the wave propagating to the right and upward. Similarly, BC and DE are parallel to wave fronts traveling to the right and downward. The angle α is the angle of incidence; the broken line FCG represents a ray reflected at C. (b) A laboratory Q-band cylindrical waveguide.*

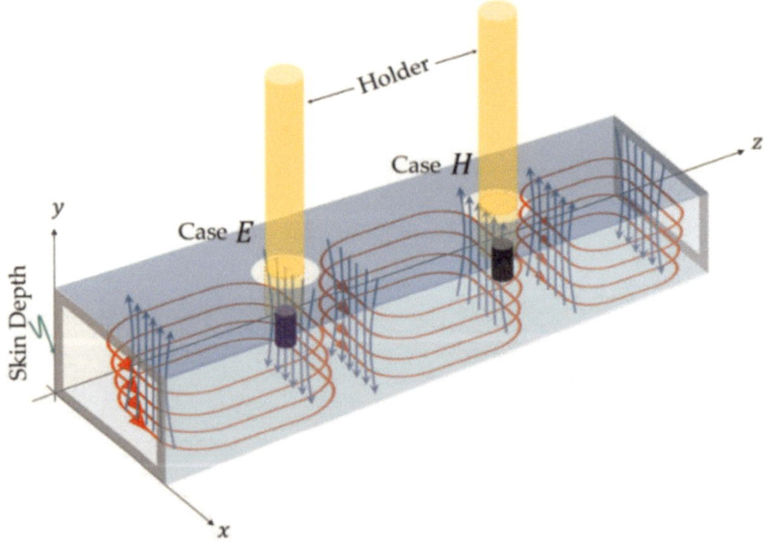


**Figure 11.**
*Fundamental idea of how to measure electrodynamic properties (ϵ, μ, σ) of matter by making it interact with microwaves inside a waveguide. Measuring transmission, reflection, or dispersion will give so much information on ϵ, μ, and/or σ of the material.*

electrodynamic properties (ϵ, μ, σ) of matter by making it interact with microwaves inside a waveguide is as follows. We just put the material specimen of interest inside the waveguide, at some place where the electric field is predominant if the electrodynamic expected response is diamagnetic, ϵ(ω), or where the magnetic field dominates, if the magnetic response, μ(ω), is to be explored. In this way the specimen will get excited electrically or magnetically, and predominantly the response would be ϵ(dielectric) or μ(magnetic) and by electron conductivity if the specimen is conductive, even if it is poor conductor (as ferrites).

To insert the specimen in the location we want, a hole is made on top of the guide and a material (dielectric, $\epsilon$; magnetic, $\mu$; and/or conductor, $\sigma$) is introduced; depending on the position of the hole **E** or **H**, the material will interact strongly with the **E** or the **H** component, see **Figure 11**, of the microwaves, and reflection, absorption, dispersion, and transmission will occur, and their measurement can be carried out.

Waveguides are used mainly to measure microwave transmission and energy transmission, T. A greater performance in the interaction of microwaves with ($\epsilon$, $\mu$, $\sigma$) materials is achieved when electromagnetic resonant cavities house the microwaves and the material to be studied.

## 5. Electromagnetic cavities

Closed metallic boxes are a particular case of a bounded space but are an important one. When the wavelength of a particular microwave (v.gr. 3 cm, 11 cm, or 8 mm) is trapped inside a box made of very good conductors (copper, silver, gold), the microwave bounces back and forth between the walls, and a pattern of standing waves is formed. The energy absorption at the walls is very small, and by virtue of this property, such a box is, really, a container of electromagnetic energy, concentrated electromagnetic energy in a closed, finite space. The same way we store a beverage in an aluminum can, we can store electromagnetic energy in a similar can (see **Figure 12**).

In electromagnetic cavities, the microwaves inside form maxima and minima at known distances, and the electromagnetic energy is stored efficiently. The main quantities that describe the electromagnetic behavior of a cavity are the standing wave electromagnetic fields **E** and **H** that form inside the cavity, its power and its figure of merit, the Q of the cavity. All these quantities are obtained from the **E** and **H** solutions to Maxwell's equations under boundary conditions at the internal metallic walls. Each component of the fields $(\mathbf{E}, \mathbf{H})$ obeys a homogeneous wave equation, as the ones given before.

They are solved by separation of variables, the boundary conditions are periodic, and the separation constants become integer numbers: $(k_1, k_2, k_3) \rightarrow (n, l, m)$ in Cartesian coordinates, or $(\chi'_{mn}, m, k)$, where $\chi'_{mn}$ are the roots of the normal Bessel functions $J'_{mn}(\chi'_{mn}, r = a)$, in spherical coordinates [28, 37, 38]. The

**Figure 12.**
*A metallic can is commonly used to contain and to store a beverage. Similarly, a metallic can is used to sustain and store electromagnetic fields in the form of standing waves.*

electrodynamic power of the EM fields inside the cavity is given by $|\mathbf{E} \times \mathbf{H}| = S = V.W/vol = Power/u.area$, and the power absorbed by a purely magnetic material is $P = \omega H_1^2 \chi''$.

If we are interested in the quantitative study of microwave dispersion, absorption, and reflection by some kind of materials, we produce standing wave patterns inside a resonant cavity, put inside the material to be studied, let the microwaves interact with it, and then measure its absorption by its reflection and its frequency shift (dispersion) by the change of the energy in the cavity (through the Q of the cavity) without and with material sample.

Examples of real microwave laboratory X-band and Q-band waveguides and cavities are as follows: the schematic representations of **E** and **H** field patterns in **Figures 9, 10** and **11** are propagated in real waveguides shown in **Figure 13**. This equipment performs high precision microwave measurements; typically the microwaves are combined with static magnetic fields to excite and saturate the magnetic specimen; hence, electromagnets are part of these equipments. The microwave circuitry includes the source, circulators, attenuators, splitters, and so forth. Some waveguides are shown to be connected through flanges to more waveguides that make 90° turns and then connected to more microwave waveguide "plumbing" until it reaches the heart of the microwave source. At the bottom end, the waveguides terminate in a rectangular cavity that hosts a specimen to be studied. Details of the microwave X-band source box are shown; an isolated open end of another waveguide is also shown. Cavities and waveguides can stand several small holes without degrading their performance; hence, several probes can be inserted. The last panel in **Figure 13** shows a rectangular cavity which is also fed in its inside with UV-vis light through an optical fiber that enters the microwave cavity space to excite simultaneously the electronic levels and the electron spins of the atoms of the specimen, the collective magnetization, and/or the domain walls of a ferromagnetic specimen. There are also dual (twin) cavities which simultaneously receive microwaves; one is empty and the other is loaded with a specimen; in real time the different absorption measurements are registered. The fact that cavities and waveguides can stand holes in their walls and insertion of different small measurement, excitation, or conducting devices, multiplies greatly the number of experimentations with $\epsilon$, $\mu$, and $\sigma$ and electronic and vibronic states that can be performed. Basically, the universal measurement in all these cases would be the reflected microwaves and from them the absorptive characteristics of the sample under study.

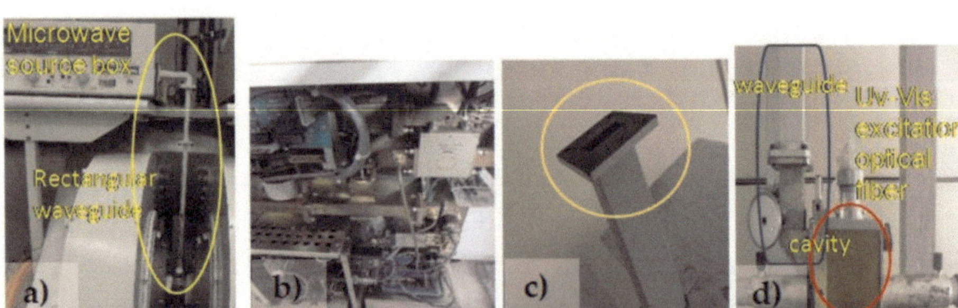


**Figure 13.**
*Common rectangular X-band waveguides and cavities and microwave source box used in laboratory. (a) typically the microwaves are combined with static* **H** *fields Electromagnets are part of these equipments, (b) the inside of a microwave source box that delivers in a precise fashion microwaves from a few microwatts up to 180 mW, (c) view of the open end of a waveguide built with a flange to connect with other waveguides, and (d) a waveguide terminated in a rectangular $TE_{011}$ cavity which is also fed with UV-Vis light through an optical fiber.*

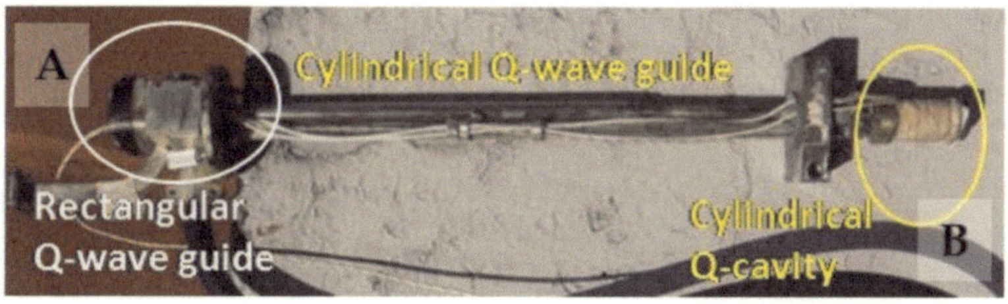


**Figure 14.**
*Multiple reflections on the inner metallic wall of a wave guide allow the propagation of microwaves long distances, A, a laboratory Q-Band cylindrical wave guide connected on the upper part to a rectangular portion of another wave guide and connected at the bottom, B, to a cylindrical resonant cavity.*

The same kind of experimental setups we just briefly described for X-band waveguides and cavities can be very well carried out at other frequencies with the appropriate K, Ku, Q, L, and S microwave equipment. The hoses that carry water and the electric cables that carry electricity can bend and give ≥90° turns, and pipes can be splitted, reduced in size, and so forth. The same with optical fibers and the same with microwave plumbing. Microwave circulators make the **E** and **H** fields go round in a circle and leave at the "aperture" of another piece of waveguide. The coupling of a waveguide with a geometry with another waveguide of another geometry is quite possible as we show in **Figure 14** for Q-band waveguides. The cylindrical waveguide connects to the right at the bottom with a cylindrical Q-band cavity, and connects to the left with a rectangular waveguide. The hole to insert a sample is at the center of the top wall (**Table 3**).

### 5.1 Circular cavity resonators

As a way of example, next we give some cylindrical frequency parameters and some standing wave patters (**Table 3**) allowed to propagate in these cavities for an

| Wave Type | $TM_{01}$ | $TM_{02}$ | $TM_{11}$ | $TE_{01}$ | $TE_{11}$ |
|---|---|---|---|---|---|
| Field distributions in cross-sectional plane, at plane of maximum transverse fields |  |  | 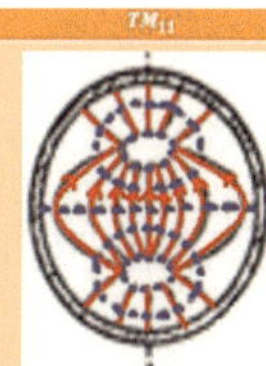 |  |  |
| Field distributions along guides | 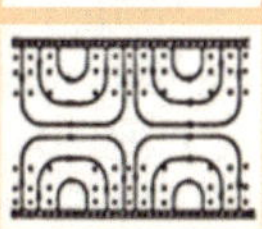 | 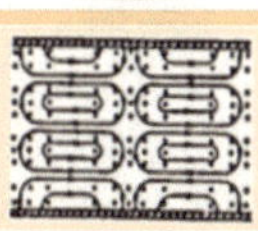 | 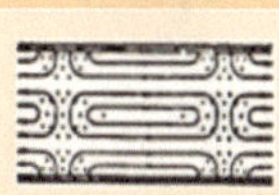 |  | 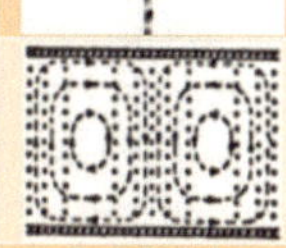 |
| Field components present | $E_{z1}\ E_{r1}\ H_\phi$ | $E_{z1}\ E_{r1}\ H_\phi$ | $E_{z1}\ E_{r1}\ E_\phi,\ H_{r1}\ H_\phi$ | $H_{z1}\ H_{r1}\ E_\phi$ | $H_{z1}\ H_{r1}\ H_\phi,\ E_{r1}\ E_\phi$ |
| $p_{nl}$ or $p'_{nl}$ | 2.405 | 5.52 | 3.83 | 3.83 | 1.84 |
| $(k_c)_{nl}$ | $\frac{2.405}{a}$ | $\frac{5.52}{a}$ | $\frac{3.83}{a}$ | $\frac{3.83}{a}$ | $\frac{1.84}{a}$ |
| $(\lambda_c)_{nl}$ | $2.51a$ | $1.14a$ | $1.64a$ | $1.64a$ | $3.41a$ |
| $(f_c)_{nl}$ | $\frac{0.393}{a\sqrt{\mu\varepsilon}}$ | $\frac{0.877}{a\sqrt{\mu\varepsilon}}$ | $\frac{0.609}{a\sqrt{\mu\varepsilon}}$ | $\frac{0.609}{a\sqrt{\mu\varepsilon}}$ | $\frac{0.293}{a\sqrt{\mu\varepsilon}}$ |
| Attenuation due to imperfect conductors | $\frac{R_s}{a_\eta}\frac{1}{\sqrt{1-\left({f_c}/{f}\right)^2}}$ | $\frac{R_s}{a_\eta}\frac{1}{\sqrt{1-\left({f_c}/{f}\right)^2}}$ | $\frac{R_s}{a_\eta}\frac{1}{\sqrt{1-\left({f_c}/{f}\right)^2}}$ | $\frac{R_s}{a_\eta}\frac{\left({f_c}/{f}\right)^2}{\sqrt{1-\left({f_c}/{f}\right)^2}}$ | $\frac{R_s}{a_\eta}\frac{1}{\sqrt{1-\left({f_c}/{f}\right)^2}}\left[\left({f_c}/{f}\right)^2+0.420\right]$ |

**Table 3.**
*Concentrate of the basic properties of some cylindrical cavities that sustain some E and H modes (patterns of standing waves). The blue dashed lines are* ***H*** *lines and the red ones are* ***E*** *lines. The second row shows how they propagate along the waveguide. The field components are shown and parameters like the cut-off frequency, the attenuation due to imperfect conductance, the cut-off wave length are also shown.*

air-filled circular cylindrical cavity resonator of radius a and length d. The resonant frequencies are

$$(f_r)_{TM_{mnp}} = \frac{1}{2\pi\sqrt{\epsilon\mu}}\sqrt{\left(\frac{\chi_{mn}}{a}\right)^2 + \left(\frac{p\pi}{d}\right)^2} \text{ and } (f_r)_{TE_{mnp}} = \frac{1}{2\pi\sqrt{\epsilon\mu}}\sqrt{\left(\frac{\chi'_{mn}}{a}\right)^2 + \left(\frac{p\pi}{d}\right)^2}$$

where the boundary condition at the lateral wall, r = a, imposes $J_{mn}(\chi_{mn}) = 0$ and the $\chi'_{mn}$ are the roots n of the m Bessel function; hence $J'_{mn}(\chi'_{mn}) = 0$

**Zeros of $J_n(x)$, $x_{np}$**

| p \ n | n = 0 | n = 1 | n = 2 |
|---|---|---|---|
| 1 | 2.405 | 3.832 | 5.136 |
| 2 | 5.520 | 7.016 | 8.417 |

**Zeros of $J'_n(x)$, $x'_{np}$**

| p \ n | n = 0 | n = 1 | n = 2 |
|---|---|---|---|
| 1 | 3.832 | 1.841 | 3.054 |
| 2 | 7.016 | 5.331 | 6.706 |

Above we show just a few roots of $J_n(\chi_{mn})$ and of $J'_n(\chi'_{mn})$. For $TE_{001}$ mode sustained in an empty cavity, m = 0, n = 0, p = 1, so $\chi'_{mn} = 3.832$, a = 3.65 cm/2 = 1.825 cm, and d = 4.38 cm with $\epsilon = \epsilon_0$ y $\mu = \mu_0$.

## 5.2 The quality factor of a resonant cavity

It is a fundamental quantity in the theory and evaluation of microwave cavities. The quality factor of the cavity is defined by

$$Q = \frac{2\pi(\text{Time} - \text{average energy stored at a resonator frequency})}{\text{Energy dissipated in one period}} \quad (35)$$

Q becomes an extremely useful parameter to measure the performance of a cavity and to make quantitative the losses in it when it is empty and/or when it is loaded with a material sample of interest. The higher the Q, the higher the quality of the metallic cavity as reservoir of electromagnetic energy. A cavity with a Q of 17,000, 33,000, or 100,000 will lose energy in a fraction of 1/17,000 or 1/33,000 or 1/100,000 of the initial energy content per cycle. So, these devices are really very good at storing microwave energy. Notice that the inverse of Q is a measure of those losses:

$Q^{-1}$ = Energy loss (absorbed, dissipated) per cycle/2π (time-avarage of energy stored at resonant frequency).

Let's call L the inverse of Q; L = $Q^{-1}$. The theory of cavities finds that there are four types, and only four types, of energy losses: (a) by Joule effect on the conducting walls and just within the skin depth, $L_\sigma$, (b) by dielectric losses if a dielectric material, $\epsilon = \epsilon' - i\epsilon''$, is introduced in the cavity, $L_\epsilon$. This means that the dielectric material absorbs microwaves by virtue of its polarized atoms/molecules, (c) by magnetic losses if a magnetic material, $\mu = \mu' - i\mu''$, is introduced in the cavity, $L_\mu$ [27–33]. This means that the magnetic material absorbs microwaves by virtue of its magnetic moments that precess with friction (damping) according to the Landau-Lifshitz equation of motion [LL], or the magnetic domain walls move back and forth trying to follow H(ω) [30], $L_\mu$, (d) any holes or apertures in the cavity, from which some microwave energy can escape, Lh. And the losses are additive, hence: $L_Q = L_\sigma + L_\epsilon + L_\mu + L_h$, in terms of $Q^{-1}$ becomes $1/Q = 1/Q_\sigma + 1/Q_\epsilon + 1/Q_\mu + 1/Q_h$. For an empty cavity: $Q = \frac{\omega W}{P_L}$ where $W = W_E + W_H$ is the

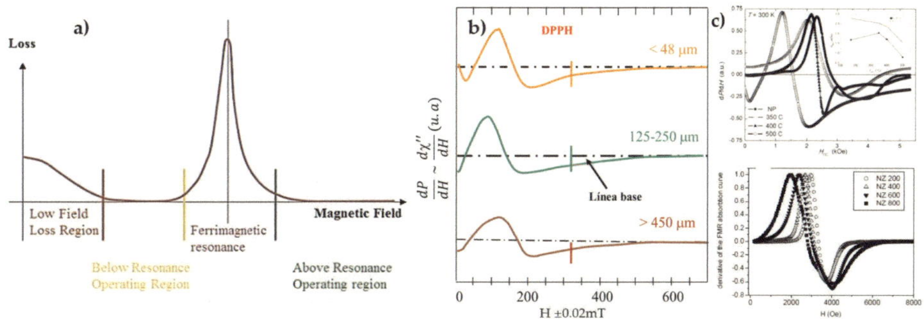


**Figure 15.**
*A variety of ferrites with different sizes and fabrication procedures absorb energy in similar ways. (a) Profile of absorption contains four distinctive regions. Only one is resonant. (b) The usual representation of the absorption is the derivative of the microwave power with respect to magnetic field. (c) Many other ferrites collapse the yellow region, and the FMR region expands to lower and higher fields.*

total electric and magnetic microwave energy and $W_E = \frac{\epsilon}{4}\int |\mathbf{E}|^2 d\tau$; $W_H = \frac{\mu}{4}\int |\mathbf{H}|^2 d\tau$. At resonance: $W = 2W_E = 2W_H$.The power loss per unit area, $L_\sigma$, is only due to the conductivity of each wall. In this case $P_{av} = \frac{1}{2}|\mathbf{J}_\sigma|^2 R_\sigma = \frac{1}{2}|\mathbf{H}|^2 R_\sigma$, where $R_\sigma$ is a superficial resistance $R_\sigma = \frac{b\rho}{\pi(D-\delta)\delta} \approx \frac{b\rho}{\pi D\delta}$, with b, length of cavity; D, diameter of cylindrical cavity; ρ, resistivity ($\rho = 1/\sigma$); σ, conductivity of cavity material; δ, skin depth $\delta = \sqrt{\frac{2\rho}{\omega\mu}} = \sqrt{\frac{2}{\omega\mu\sigma}}$; ν, electromagnetic wave frequency; and $J_s = |\mathbf{J_s}|$, AC current density generated by microwaves inside the cavity wall, generating a power loss $P_S$. Now, what follows has been found experimentally [39, 40]. If an extra conductor as a wire or a conducting film is introduced in the cavity, a new loss term due to $J_{ex}$ (AC current density generated by microwaves within the skin depth of the extra conductor) appears, $P_{ex}$. Both terms are of the same type; hence, $P_{\sigma total} = P_\sigma + P_{ex}$, and both power loss integrals are of the same type. $P_\sigma = \oint P_{av}\, ds$, and the new Q is: $Q = \frac{\omega W}{P_\sigma + Px} = \frac{2\pi\nu W}{\oint P_{av}\, ds}$. And so, $1/Q = 1/Q_\sigma + 1/Q_{ex}$. If we continue adding lossy objects inside a cavity, more power loss terms appear and the total loss would be the sum of each loss, $P_{total} = \sum P_i$, where $P_i = P_\sigma + P_\epsilon + P_\mu + P_h$. This analysis on the Q of a loaded cavity and its losses is very powerful to understand what mechanisms are responsible for the total microwave absorption that a magneto conductor exhibits.

### 5.3 Examples of experimental measurements of microwave absorption by different magnetic and/or conducting materials

The first example is microwave ferrites. Their name clearly indicates the main function they have and have been studied with microwaves since their very invention. Microwave ferrites are crucial elements in microwave measurement equipment itself and in a pleyade of different microwave devices [17]. Its ability to absorb greatly microwaves under very specific circumstances and do not absorb them under other set of circumstances makes these materials highly controllable, and that is what engineering requires [11].

As microwave device it is desirable to have wide yellow and green regions in **Figure 15** for passive circulator and isolator operation [11, 24]. Ferrites absorb microwave energy in a resonant fashion and under nonresonant conditions, making these responses a very versatile and manageable material. It is very cheap and easy to fabricate [41].

## 6. Resonant absorption of microwaves

What is microwave energy absorption in a resonant fashion? The phenomenon is really ferromagnetic resonance (FMR). What is the role of microwaves in the ferromagnetic resonance phenomenon? A brief description follows; the most common measurements of the ferrite absorption performance or profile are carried out in equipment as the one shown in **Figure 13**. In addition to the microwave excitation of the ferrite inside the cavity, an extra static magnetic field, $H_0$, is applied to the ferrite through the magnet poles of the electromagnet also shown in **Figure 13a**. This is why the cavity is seen located at the center of the magnet poles. This field serves to simplify the magnetic structure of domains of the ferrite, and when 250 mT (2500 Oersters) or more are applied, the domain structure has disappeared, and the material becomes magnetically saturated, and the whole sample has the magnetization value $M_s$, and this $M_s$ as a whole interacts with the microwave magnetic field and absorbs its energy greatly in the form of $\hbar\omega = g\beta H_0$, in which $\hbar\omega$ is the energy of a quantum of the microwave field and the right side $g\beta H_0$ is the magnetic energy splitting of two consecutive magnetic energy levels. This is the well-known Zemann effect, where $\beta = \hbar e/2m_e$ is Bohr magneton and g is the spectroscopic factor (for ferromagnets and ferrimagnets g is close to 2.00 but always larger). To have $\hbar\omega = g\beta H_0$ means that a stimulated transition between two contiguous energy levels, $\Delta E = g\beta H_0$, is taking place and the energy is provided by photons, $h\nu$, with v exactly in the microwave region, of the magnetic component of the microwaves. This is called the resonance condition; it is fully quantum and was discovered without knowing what it was in 1946 by R. Griffiths [34] and explained fully 1 year later by Kittel [32]. When this absorption fullfils the Kittel condition, $h\nu = g\beta H_0$, it is resonant absorption of energy (no more, no less, just exactly the energy content in a microwave photon $h\nu$), meaning that resonant absorption of microwave (photon) energy is performed by the magnetization of a ferromagnetic specimen. When the atomic magnetic moments or uncoupled electron spins are not governed by the strong magnetic exchange interactions, they do not behave collectively, as a unit, and they behave individually. Such is the case of paramagnetic substances. Each atomic magnetic moment, $\mathbf{m}_i$, or electron "spin only", **S**, can and do absorb microwaves individually obeying also the resonance energy equation, $h\nu = g\beta H_0$. In these cases the phenomenon is called electron paramagnetic resonance (EPR) or electron spin resonance (ESR). In any of these cases, energy from the microwaves is absorbed resonantly and very efficiently. So, FMR, EPR, and ESR are techniques that measure very accurately the absorption of microwaves (quanta, $h\nu$) in the presence of a static magnetic field, $H_0$, of a magnetic sample located in a microwave cavity.

## 7. Nonresonant absorption of microwaves

When the microwave experimental setup is as described but the equality $h\nu = g\beta H_0$ is not fullfiled and absorption of microwaves is still registered, then we have a nonresonant absorption of microwaves, and other dissipative process dynamics are taking place. For example, domain walls can be made to oscillate with an external field, and the motion is dissipative, or some conduction "currents" can be operating. So, the same equipment and experimental setup can measure resonant and nonresonant absorption of microwaves. Measurements of this kind, carried out in equipment as in **Figure 13a** on ferrites fabricated with different methods and with different compositions [41], give very frequent absorption profiles as the ones shown in **Figure 15**. This contains nonresonant and resonant absorption of microwaves. This kind of absorption profile has been known for many years. Here we want to demonstrate that resonant and nonresonant absorption of microwaves coexists in just one measurement that is capturing different microscopic absorption mechanisms at different values of $H_0$. The information obtained this way is very rich. For passive microwave circulators and isolators, it is highly convenient that the regions below resonance (B/R, yellow) and above resonance (A/R, green) be as wide as possible since no absorption is demanded. The widening of these yellow and green regions is a continuous search by modifying ferrite fabrication parameters and continuously measuring this kind of microwave absorption. Yet, many ferrites do the contrary and absorb in all regions of $H_0$. On the other hand, the maximum possible absorption is required in order to sensing it from a distance. In a sense a kind of sink is desirable, like an antenna that works by absorbing greatly microwaves. Much the same way radar works. In order to develop potential applications as the one illustrated in **Figure 3**, "sensors" that absorb greatly microwaves in preferred directions are required. Some promising materials are $Fe_{79}B_{10}Si_{11}$ glass-covered amorphous-conducting magnetic microwires (simply FeBSi wires) because they have shown great capacity to absorb microwaves at X-band in an anisotropic fashion [42]. The proposed application in **Figure 3** demands a great global absorption of microwaves in order to detect reflected microwaves from implanted magnetic microwires (glass-covered for them to be biologically inert) in patients that have undergone some kind of orthopedic surgery at the level of knee, shoulder, vertebra, hip, and so on. The microwires are implanted with some specific orientation, and as recovery develops and bone grows, or fractures heal, the microwires would move when pushed by the new processes taking place. Those changes are expected to be informative to the surgeons. The idea of the detection is quite similar as how radar detects moving or static "objects" at a distance. The same idea is used in how the laser gun works detecting a speeding vehicle. A good level of reflected, or perturbed. microwave "signal" coming back to the transducer is required. In laboratory models, FeBSi wires have shown great microwave absorption at some particular orientations. Experiments are carried out with the wires inside microwave cavities in equipment as the one shown in **Figure 13**. Conditions are established for FMR absorption because it is the maximum possible; hence in addition to the X-band microwaves fed to the resonant cavity, an extra static magnetic field is also applied. The physical interactions are as described for ferrites, except that these are amorphous and no long-range order exists and the strong

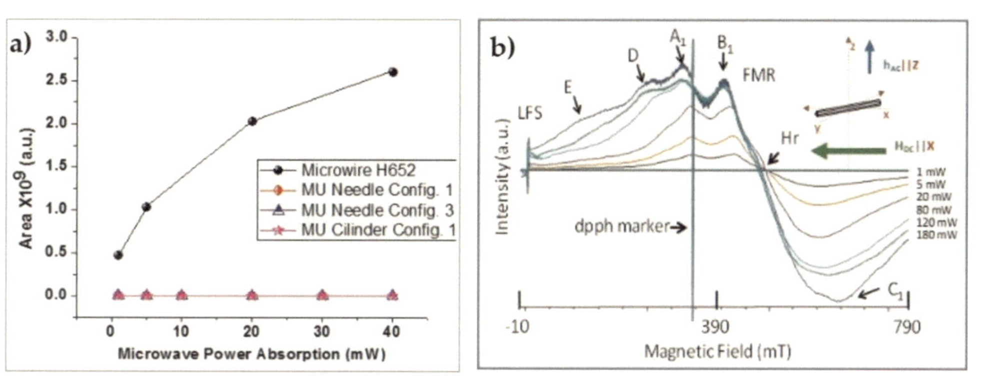


**Figure 16.**
*FMR absorption of microwaves by glass-coated amorphous-conducting ferromagnetic microwires. (a) First derivative of absorption intensity as function of delivered microwave power and (b) the total absorption is a fast-increasing function of power. The Abs% grows to over 1000% for P = 180 mW. The measured absorption for the military material "mu" is shown for comparison, it does not show such effect [42].*

crystalline anisotropy does not exist and $H_{int}$ that goes in $H_{eff}$ is different from the $H_{int}$ present in ferrites. A typical absorption profile is shown in **Figure 16**. Here the quantity of interest is the integral of the absorption curve since it gives directly the total microwave power absorbed by the sample. The higher the integral, the better in order to use it as sensor-detector of microwaves. The total absorption resulted in a fast-increasing monotonic function of power. The Abs% grows to over 1000% for P = 180 mW. The measured absorption for the military material "mu" is shown for comparison; it does not show such effect. This is a kind of amplification effect [42].

In, yet, another application, the cavity microwave magnetic field near an extra inserted conducting perturbation is greatly enhanced. Nanomagnets and micromagnets could require for their study an enhancement of the fields they experience inside a resonant cavity. This could be achieved by introducing an extra conductor (wire) in the cavity in order to expose its free electrons to the microwave electric field in the cavity. Induced currents in the conductor, of the same frequency of the microwaves, would produce an extra magnetic field $H^+$ in some small regions, $\gamma$, very close to the extra wires, $w^+$. Placing a micron- or nano-sized sample, $\eta$, in region $\gamma$ of increased field $H_{incr} = H_0 + H_{int} + H^+$ could produce an amplified ferromagnetic resonance absorption, since now $\hbar\omega = g\beta H_0$ is fulfilled as before, but more microwave effective power is absorbed by a micro- or nano-magnetic material with a not so large total magnetic moment, M, placed at the $\gamma$ region. This was proven experimentally by R. Rodbell in 1952 [39, 40] more than half a century ago; we consider it a classic of deep understanding of electrodynamics in cavities and a good example of how to use them in novel ways. Present-day microwave experiments of these kinds can be performed on ferromagnetic resonance equipment that looks like the one shown in **Figure 13**.

One interesting result obtained by Rodbell is that the microwave magnetic field strength near the surface of a conducting rod may be easily made to exceed the maximum magnetic field strength existing in the unperturbed cavity at the same

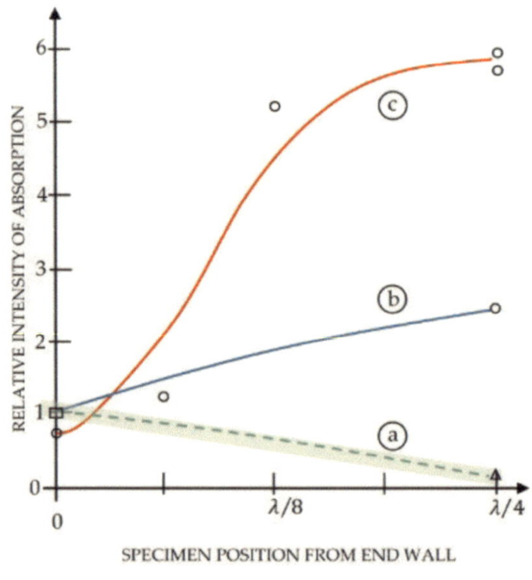


**Figure 17.**
*The intensity of the ferromagnetic resonance measured for a sample of $MnFe_2O$ powder as it depends upon position of the sample in a rectangular $TE_{011}$ X-band microwave cavity. The incident power is 50 mW. (a) FMR with the ferrite alone, (b) ferrite near a cooper wire, (c) ferrite powder glued to the copper wire. The FMR absorption increase amplification is more than 1000X.*

incident power level. This would appear to be a useful means of effectively coupling microwave energy into a magnetic specimen. The experiment described by Rodbell is the measurement of the relative intensity of the microwave magnetic resonance absorption of an $MnFe_2O_2$ powder specimen (manganese ferrite) as it depends upon position within a rectangular (TE) microwave cavity at $\approx 9$ kMc/s, **Figure 17**. The powder specimen is cemented onto the outside surface of a quartz capillary tube that is 3 mm long and of 0.25 mmo.d. The capillary is attached to a quartz post so that it can be positioned along the central "$E$" plane of the cavity with the capillary axis along the microwave **E** field. The total resonance absorption here is composed of contributions from many, essentially isolated, randomly oriented particles of the ferrite powder; the line width of the composite absorption is about 1000 Oe. The resonance absorption is used here as an indication of the square of the microwave magnetic field strength averaged over the sample. This is the variation expected for a magnetic resonance absorption that is driven by the usual microwave magnetic field strength for this cavity geometry. The experiment is now repeated after introducing into the capillary tube a bare copper wire that is 3 mm long and of 0.025 mmdiam. The cavity coupling and incident microwave power are constants of the experiment. The small variation of the cavity Q with dc magnetic field is used as a measure of the magnetic resonance absorption in the standard way.

The results indicate that the electric field "drives" the absorption; that is, the microwave electric field is locally perturbed and gives rise to a locally large magnetic field. Further confirmation is found in curve (c) of **Figure 17** which displays the result of an experiment in which a copper wire of the same size as in (b) is coated directly (no intervening quartz capillary) with approximately the same amount of the ferrite ($MnFe_2O_4$) powder. The increased absorption, here relative to (b), is interpreted to be the result of the larger microwave magnetic field that occurs closer to the perturbing conductor.

## 8. Conclusions

An overview of the universality of the microwaves in the universe and in the modern technology world was given. Maxwell's equations are placed in the center of the electrodynamics universe and in particular in all the technological applications that were mentioned: microwaves in open spaces as in radar, Wi-Fi, guided microwaves, and microwaves in closed resonant cavities with very good conducting walls. Solutions of Maxwell's equations are given in tables, and the phenomena of reflection, refraction, and absorption are shown to be universal. The Snell and Fresnel reflection and refraction equations are given. The fundamental physics of propagation is given, and the main features of electromagnetism in resonant cavities were described. Real waveguides and cavities and microwave equipment were presented, and general ideas of their use for research were also given. Few applications of absorption measurements under ferromagnetic resonance and nonresonant conditions for ferrites, amorphous microwires, and conductors with magnetism were given.

## Author details

Rafael Zamorano Ulloa[1]*, Ma. Guadalupe Hernandez Santiago[2]
and Veronica L. Villegas Rueda[3]

1 Physics Department, Superior School of Physics and Mathematics, National Polytechnic Institute, Mexico City, Mexico

2 Mathematics Department, Faculty of Superior Studies FES Cuautitlan, UNAM, State of Mexico, Mexico

3 Basic Sciences Department, Professional Interdisciplinary Unit of Engineering and Advanced Technology, IPN, Mexico City, Mexico

*Address all correspondence to: davozam@yahoo.com

## References

[1] Penzias AA, Wilson RW. A measurement of excess antenna temperature at 4080 Mc/s. Astrophysical Journal Letters. 1965;**142**: 419-421. DOI: 10.1086/148307

[2] Penzias AA, Wilson RW. A measurement of the flux density of CAS A at 4080 Mc/s. Astrophysical Journal Letters. 1965;**142**:1149-1154. DOI: 10.1086/148384

[3] National Aeronautics and Space Administration. Tests of Big Bang: The CMB [Internet]. Available from: https://wmap.gsfc.nasa.gov/universe/bb_tests_cmb.html [Accessed: 04-December-2018]

[4] NASA Science. Tour of the electromagnetic spectrum. Microwaves [Internet]. https://science.nasa.gov/ems/06_microwaves [Accessed: 04-December-2018]

[5] Electronic Design Innovation Conference 2018. EDI CON USA [Internet]. Available from: http://www.ediconusa.com/ [Accessed: 04-December-2018]

[6] Glyn S, Smith R, Devlin L, Dearn A, Pearson G. Design of the single chip front-end module for 28 GHz 5G. Microwave Journal. 2018;**61**(4):22-35

[7] Calif I. Defense opportunities and challenges in 2019. Microwave Journal. 2018;**61**(11):22-32

[8] Panina LV et al. Giant magneto-impedance and magneto-inductive effects in amorphous alloys. Journal of Applied Physics. 1994;**76**(10):6198. DOI: 10.1063/1.358310

[9] Jeol. Instructions ESR Data System. Tokio. Japan: Ed. Jeol LTD; 1992

[10] Bruker. EPR Resonators. Ed. BrukerBiospin, [Internet]. Available from: https://www.bruker.com/es/products/mr/epr.html [Accessed: 04-December-2018]

[11] Özgür U, Alivov Y, Morkoc HA. Microwave ferrite, part 1: Fundamental properties. Journal of Materials Science: Materials in Electronics. 2009;**20**(9): 789-834

[12] Pardavi-Horvath M. Microwave applications of soft ferrites. Journal of Magnetims and Magnetic Materials. 2000;**215-216**:171-183

[13] News and Features. Krohne Group. [Internet]. Available from: https://krohne.com/en/ [Accessed: 04-December-2018]

[14] Shenhreen A, Manoh K, Veerakumar V, Bijoy KK. Microwave monolithic filter and phase shifther using magnetic nanostructures. American Institute of Physics. 2018;**8**: 056624

[15] Huai-Wu ZT, Li J, Hua S, Chuan ZT, Yang L, Liang ZZ. Development and application of ferrite materials for low temperature co-fired ceramic technology. Chinese Physics B. 2013;**22** (11):117504

[16] Santre A, Nasr I, Kim L. Reinventing radar: The power of 4D sensing. Microwave Journal. 2018;**61**(12):22-37

[17] Goldman A. Modern Ferrite Technology. 2nd ed. Springer; 2006. p. 218

[18] IEEE MTT-S International Microwave Symposium [Internet]. Available from: https://ims2018.org/ [Accessed: 04-December-2018]

[19] ESA and the Plank Collaboration. Plank Maps the Microwave Background [Internet]. Available from: https://apod.

nasa.gov/apod/ap180722.html [Accessed: 04-December-2018]

[20] DeTomasi S. Navigating the 5G NR standards. Microwave Journal. 2018;**61**(12):72-83

[21] NASA Science. Tour of the electromagnetic spectrum. Microwaves [Internet]. https://science.nasa.gov/ems/06_microwaves [Accessed: 04-December-2018]

[22] Microwave Energize Medical Applications [Internet]. Available from: https://www.mwrf.com/medical/medical-applications-seek-wireless-solutions [Accessed: 04-December-2018]

[23] Graduad School of frontier Science, University of Tokyo. Success in Thrust Enhancement of Microwave Rocket [Internet]. Available from: http://www.k.u-tokyo.ac.jp/news/20100105press-e.html [Accessed: 04-December-2018]

[24] Hinsaw G et al. Nine years Wilkinson Microwave Anisotropy Probe (WMAP) observations: Cosmological parameters results. The Astrophysical Journal Supplement Series. 2013;**208**(19):25. DOI: 10.1088/0067-0049/208/2/19

[25] National Aeronautics and Space Administration. Wilkinson Microwave Anisptropy Probe (WMAP) [Internet]. Available from: https://map.gsfc.nasa.gov/ [Accessed: 04-December-2018]

[26] Spliter LG et al. A repeating fast radio burst. Nature. 2016;**531**(7593): 202-205

[27] Lorrain P, Corson D. Electromagnetic Fields and Waves. 2nd ed. W H Freeman and Company; 1970. 696 p. ISBN: 0-7167-0331-9

[28] Jackson JD. Classical Electrodynamics. John Wiley & Sons, Inc.; 1962. 656 p

[29] Feynman RP, Leighton RB, Sands M. The Feynman Lectures on Physics. Mainly Electromagnetism and Matter. Vol. II. The New Millennium Ed. Basic Books; 2010. 566 p. ISBN: 978-0-465-07998-8

[30] Landau LD, Lifshitz EM. Electrodynamics of Continous Media. Course of Theoretical Physics. Vol 8. 2nd ed. Pergamon Press; 1984. 457 p

[31] Morrish AH. The Physical Principles of Magnetism. John Wiley & Sons, Inc.; 1965. 680 p

[32] Kittel C. Introduction to Solid State Physics. John Wiley & Sons, Inc.; 1953. 408 p

[33] Griffiths DJ. Introduction to Electrodynamics. 3rd ed. Prentice Hall; 1999. 576 p

[34] Griffiths JHE. Anomalous high-frequency resistance of ferromagnetic metals. Nature. 1946;**158**(4019):670-671. DOI: 10.1038/158670a0

[35] Protheroe RJ, Johnson PA. Propagation of ultra high energy protons over cosmological distances and implications for topological defects models. Astoparticle Physics. 1996;**4**(3): 253-269. DOI: 10.1016/0927-6505(95)00039-9

[36] Espinosa JR, Racco D, Riotto A. Cosmological signature of the standard model higgs vacuum instability: Primordial black holes as dark matter. Physical Review Letters. 2018;**120**: 121301. DOI: 10.1103/PhysRevLett.120.121301

[37] Slatter JC. Microwave electronics. Reviews Modern Physics. 1946;**18**:144. DOI: 10.1103/RevModPhys.18.441

[38] Straton JA. Electromagnetic Theory. International Series in Physics. 1st ed. McGraw Hill Book Company Inc.; 1941. 631 p

[39] Rodbell DS. Ferromagnetic resonance of iron whisker crystals. Journal of Applied Physics. 1959;**30**:5178

[40] Rodbell DS. Microwave magnetic field near a conducting perturbation. Journal of Applied Physics. 1959;**30**:1845

[41] Hernández Santiago MG. Estudio FMR en ferritas [Mater's Thesis]. México, DF: ESFM-IPN;. 2019

[42] Villegas Rueda VL. Estudio FMR de microhilos ferromagnéticos de composición $Fe_{79}B_{11}Si_{10}$ [Master's thesis]. México, DF: ESFM-IPN; 2008

Chapter 7

# Electromagnetic Field Interaction with Metamaterials

*Mohammed M. Bait-Suwailam*

## Abstract

It is well known that constitutive parameters, namely, the electrical permittivity, $\varepsilon$, and the magnetic permeability, $\mu$, in a medium determine the response and reaction of such medium or material when exposed to external time-varying electromagnetic fields. Furthermore, most materials are lossy and dispersive, that is, both permittivity and permeability are complex and frequency-dependent. Interestingly, by controlling the sign of real parts of $\varepsilon$ and $\mu$ in a medium, unique electromagnetic properties can be achieved that are not readily available in nature. Recently, subwavelength composite engineered structures, also known as *metamaterials*, have evolved in many engineering and optical applications, due to their unique electromagnetic properties that are not found in nature, including but not limited to negative refractive index, backward wave propagation, subwavelength focusing and super lenses, and invisibility cloaking. The main aims of this chapter are to provide an overview of electromagnetic field behavior and interaction with metamaterials and to explore such behavior in various metamaterials both analytically and numerically.

**Keywords:** double negative medium, electromagnetic waves, metamaterials, plane wave, single-negative medium

## 1. Introduction

Electromagnetic field is a physical behavior that is produced in a space due to time-varying electric charges and represents the interaction between electric and magnetic fields. Unlike static charges that can only produce static electric fields in space, time-varying electric charges are one of sources for the rise of magnetic fields, which in turn produce time-varying electric fields. This is summarized in the four time-varying Maxwell's equations given in differential form:

$$\nabla.\mathbf{E} = \rho_v(t)/\varepsilon \tag{1}$$

$$\nabla.\mathbf{B} = 0 \tag{2}$$

$$\nabla \times \mathbf{E} = -\mu\frac{\partial \mathbf{H}}{\partial t} \tag{3}$$

$$\nabla \times \mathbf{B} = \boldsymbol{J}(t) + \varepsilon\frac{\partial \mathbf{E}}{\partial t} \tag{4}$$

where $\rho_v$ is time-varying volume charge density, $\varepsilon$ and $\mu$ are the electric permittivity and magnetic permeability, respectively, $\boldsymbol{J}$ is the time-varying electric current



IntechOpen

density in a medium, **D** and **B** are time-varying electric and magnetic flux densities, respectively, and **E** and **H** are time-varying electric and magnetic field intensities, respectively.

In 1864, James Maxwell showed through Eqs. (1)–(4) that oscillating electric and magnetic fields give rise to electromagnetic waves that travel at the speed of light in free space [1], which also implies that light is electromagnetic in nature. By taking the curl of Eqs. (3) and (4), it is also straightforward to show that electromagnetic wave propagation can exist.

In a medium, there are two main quantities, also known as the constitutive parameters, namely, electric permittivity, $\varepsilon$, and the magnetic permeability, $\mu$, in addition to the conductivity, $\sigma$, that determine the nature of electromagnetic wave and its behavior in such a medium. In other words, the aforementioned parameters along with the boundary conditions in a medium determine uniquely the response of such medium to an incoming electromagnetic wave. This is also summarized through two equations, given below, that describe the relationship between electric and magnetic field quantities in a simple linear and isotropic medium:

$$\mathbf{D} = \varepsilon\mathbf{E} \tag{5}$$

$$\mathbf{B} = \mu\mathbf{H} \tag{6}$$

where in Eqs. (5)–(6), both $\varepsilon$ and $\mu$ in a lossy dispersive medium are commonly complex and frequency-dependent and are real quantities in a lossless isotropic medium. From such relations, Eqs. (1)–(6), important parameters, such as the wavenumber, *k,* the refractive index, *n*, and the intrinsic wave impedance, $\eta$, in a medium can be determined, which are given respectively as:

$$\mathrm{k} = \omega\sqrt{\mu\varepsilon} \tag{7}$$

$$\eta = \sqrt{\mu/\varepsilon} \tag{8}$$

$$n = \sqrt{\mu_r\varepsilon_r} \tag{9}$$

where $\omega = 2\pi f$ is the radian frequency (in rad/sec), $f$ is the frequency (in Hz), and $\mu_r = \mu/\mu_0$ and $\varepsilon_r = \varepsilon/\varepsilon_0$ are the relative permeability and permittivity, respectively, while $\mu_0$ and $\varepsilon_0$ are the free-space permeability and permittivity, respectively.

## 2. Overview of metamaterials and their realizations

**Figure 1** depicts a general overview of possible materials based on their constitutive parameters: the electric permittivity and the magnetic permeability values. The aforementioned constitutive parameters are in principle complex, and their signs are based on the signs of their real parts, while their imaginary parts indicate the presence of electric or magnetic losses, respectively. While in naturally occurring materials, both real parts of the permittivity and permeability are positive (i.e., >0); it is possible that either one of the real parts of the constitutive parameters or even both have negative values. In the second quadrant, while permeability is above zero, the permittivity is below zero (negative), which can be termed as a single-negative (SNG) or $\varepsilon$-negative (ENG) medium. Similarly, when a medium possesses negative permeability value, while its permittivity is positive, this is also termed as an SNG medium or $\mu$-NG (MNG) medium, where $\mu < 0$ in this type of material, as shown in the third quadrant in **Figure 1**. An interesting medium is the case when both real parts of permittivity and permeability are negative (i.e., the third quadrant in **Figure 1**). This is termed as double negative (DNG) medium or left-handed

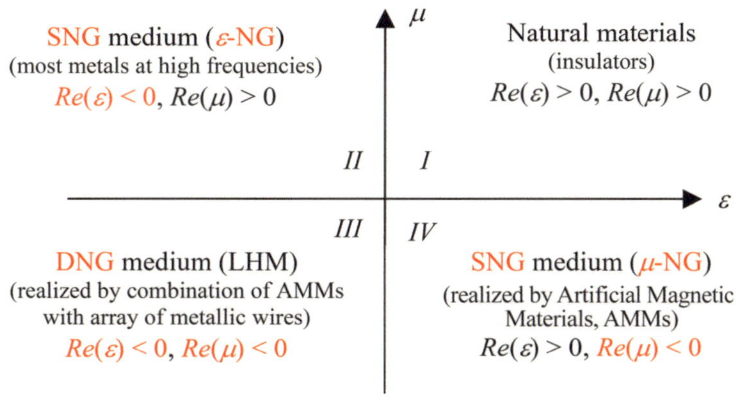


**Figure 1.**
*Classification of materials, based on their constitutive parameters, ε and μ.*

medium (LHM), due to its unique resultant electromagnetic features, like negative refraction and negative phase velocity, as it follows a left-handed system rule. In summary, metamaterials have three classes, depending on the signs of their constitutive parameters: ENG, MNG, and DNG.

Practically, natural SNG media are available that possess ENG response, for instance, metals at visible and near-ultraviolet regime. However, at much lower frequencies, one commonly adopted realization of SNG medium is the periodic arrangement of metallic wires, which results in possessing negative effective permittivity below the plasma frequency of metallic wires or rods [2]. It is instructive to mention here that naturally occurring materials with permeability values below zero are not yet available in nature, especially within the radio frequency/microwave regime. However, such response can be obtained through engineered arrangement of metallic inclusions printed on a dielectric medium [3], as it will be discussed further later on.

After the seminal work of Veselago in [4], where he investigated mathematically the possibility of electromagnetic wave propagation through materials with both negative permittivity and permeability values, the word "metamaterials" evolved, which refers to what is beyond naturally occurring materials. Metamaterials can be defined as artificially engineered structures that have electromagnetic properties not yet readily available in nature. Such artificial composite structures are realized in one way by periodically patterning metallic resonant inclusions in a host medium, i.e., dielectric or magneto-dielectric material, either in a symmetric or nonsymmetric fashion. When exposed to an electromagnetic field, the metamaterials alter the electromagnetic properties of the new host medium due to mainly the inclusions' response and features. **Figure 2** depicts a general view of one possible realization of a metamaterial structure.

Tremendous efforts had been put forward in the past with the goal to provide efficient numerical means for the retrieval of constitutive parameters of arrays of metamaterials in order to advance the design and characterization of metamaterials [5–7]. Such numerical retrievals provided engineering and physical means in replacing local electromagnetic response details of individual metamaterials elements with averaged or homogenized values for the effective electric permittivity, $\varepsilon_{eff}$, and effective magnetic permeability, $\mu_{eff}$. As a matter of fact, this retrieval approach is a direct translation of the characterization of natural media, which consist of atoms and molecules with their dimensions that are much smaller in magnitude than the wavelength. The electromagnetic wave response and propagation within the effective metamaterial medium can then be fundamentally

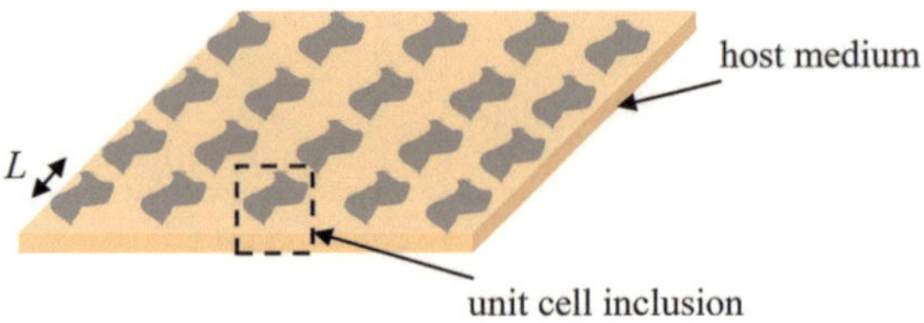


**Figure 2.**
*General sketch of a metamaterial composite structure.*

described using constitutive parameters along with Maxwell's equations. In principle, this effective response would be permissible if the unit cell dimension is sufficiently small enough or a fraction of an operating wavelength [5], say, for example, $L$, as shown in **Figure 2**, satisfies the relation below:

$$L \ll \lambda \tag{10}$$

where $L$ is the unit cell dimension and $\lambda$ is the operating wavelength of the incoming electromagnetic field. When the condition in Eq. (10) holds, quasi-static behavior for the artificial metamaterials can be applied, in which an equivalent resonant circuit, composed of *resistor-inductor-capacitor (RLC) elements*, is permissible to use in order to provide qualitative description of the physical behavior of the artificial materials [8].

## 2.1 Realization of artificial μ-negative (MNG) medium

Among engineered materials with negative magnetic permeability, artificial magnetic materials (AMMs) have been the subject of interest for many years. This is due to their unique features, including low cost ease of integration with radio frequency/microwave circuits, and the possibility of synthesizing magnetic permeability to certain magnetization and polarization levels at the frequency of interest. This is in contrast to ordinary magnetic materials, like ferrite composites, that are limited in their magnetization levels and as well as suffer from magnetic losses at microwave frequency regime [9].

The idea of creating magnetic materials from conductors was first proposed by Schelkunoff [10]. A wide variety of artificial magnetic inclusions have been proposed and implemented in the literature. One of the popular and widely applied artificial magnetic materials is the split-ring resonator (SRR). Pendry et al. [3] used concentric metallic rings in order to provide further enhancement of the magnetic properties of the rings. The SRR, as shown in **Figure 3a**, consists of two concentric circular (i.e., edge-coupled) metallic rings printed in a host dielectric medium, with splits at opposite ends of the rings. Another form of "SRR-based" AMM is realized by placing the two split-rings in opposite sides (broadside-coupled) within the host medium [11, 12], which can provide two advantages: firstly, the effects of bianisotropy, or cross polarization, are eliminated due to the broadside nature of the metallic rings (see **Figure 3c**), and secondly, there is additional capacitive coupling to the composite structure, hence achieving stronger resonance behavior [8, 11, 12], as shown in **Figure 3b**. Other forms of resonant metallic inclusions, like spiral, omega, Hilbert, can also be adopted to achieve artificial magnetic media, as shown in **Figure 3**. Significant miniaturization factors can be achieved using either spiral- or Hilbert-type resonators [13, 14].

The physical principle of operation behind the artificial magnetic materials, as shown in **Figure 3**, is almost the same. Let us consider the subwavelength resonant inclusion in **Figure 3c** and assume that it occupies an infinite space with large number

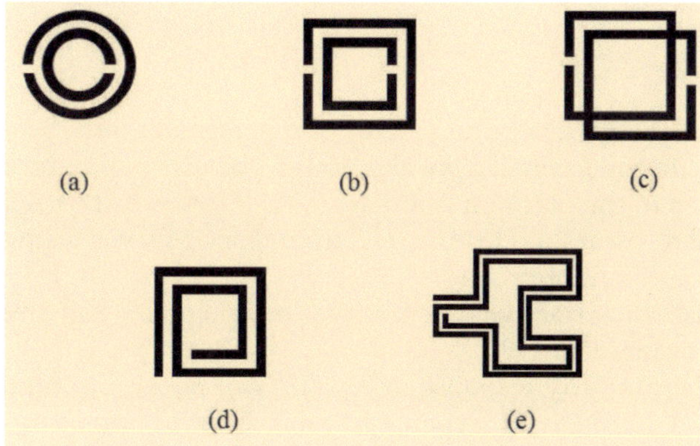


**Figure 3.**
*Possible resonant metallic inclusions for synthesizing artificial magnetic materials, (a) edge-coupled circular SRR, (b) edge-coupled square SRR, (c) broadside-coupled SRR, (d) spiral resonator, and (e) Hilbert resonator.*

of periodicity in two- or three-dimensional planes (i.e., periodicity implies here repetition factor that is much smaller than λ). Upon an excitation of an external magnetic field, which is orthogonal to the paper plane, to such an infinitely large and homogenized artificial structure, the external magnetic field induces an electromotive force in the inclusions, which in turn results in a circulating effective current flowing around the inclusions. Upon such excitation, a general form for the effective magnetic permeability of any of the AMM structures as shown in **Figure 3** can then be expressed as

$$\mu_{eff} = 1 - \frac{K\, j\omega L_{eff}}{Z_{inc} + j\omega L_{eff}} \tag{11}$$

where $K$ is a normalized fractional surface area that is enclosed by the AMM inclusion and $L_{eff}$ is the effective inductance of the AMM, which is given by

$$L_{eff} = \frac{\mu_0 S}{p} \tag{12}$$

where $S$ is the surface area of the AMM and $p$ is the periodicity of the AMM inclusion that mimics an infinitely large AMM structure. The parameter $Z_{inc}$ in Eq. (11) consists of two parts: $R_{eff}$, which represents the encountered ohmic losses due to the finite conductivity of the metallic rings within the AMM inclusions, and $C_{eff}$, which is the mutual capacitive effect due to the close proximity of the AMM metallic rings/strips. Comprehensive analytical modeling approaches can be found in [3, 8, 12–14].

## 2.2 Realization of artificial ε-negative (ENG) medium

As discussed in the previous section, the realization of metallic resonant inclusions patterned in a homogenized host medium had made it possible to synthesize magnetic permeability in the microwave and optical regimes [3]. Similarly, it is possible to engineer the permittivity of a bulk medium by facilitating patterned metallic inclusions. In solid metals, negative permittivity response commonly occurs at the visible and near-ultraviolet regime, due to the entire oscillation of plasmons [15]. Pendry et al. showed that an array of thin rods or wires arranged in a cubical lattice can indeed exhibit negative effective permittivity response at the microwave regime given by the Drude function [2]:

$$\varepsilon(\omega) = 1 - \frac{\omega_p^2}{\omega(\omega + j\Gamma)} \tag{13}$$

where $\Gamma$ is the energy dissipation factor of the plasmon into the system (i.e., damping factor). In solid metals, like aluminum, the dissipation factor, $\Gamma$, is usually small as compared to the plasmon frequency, $\omega_p$. If losses were neglected, (i.e., $\Gamma \approx 0$), it is evident from Eq. (13) that electromagnetic waves below the plasma frequency ($\omega_p > \omega_p$) cannot propagate, since $\varepsilon < 0$ ($\mu$ here >0). It is also evident from Eq. (13) that the refractive index, $n$, will be imaginary and a wave in such a medium will be evanescent.

The electric permittivity response in Eq. (13) can be used to represent the effective electrical permittivity response of a synthesized homogeneous medium comprising an array of very thin metallic wires. Note that in the realization of the composite periodic lattice of wires, wires' diameter is essentially much smaller than the operating wavelength in order to mimic an effective homogenized negative permittivity media from such periodic metal rods. The term $\omega_p$ represents the plasma frequency for metals and can be expressed in terms of the electron properties by following relation [2]:

$$\omega_p^2 = \frac{nq^2}{\varepsilon_0 m_e} \tag{14}$$

where $q$ is the electron charge, $\varepsilon_0$ is the free-space permittivity, and $n$ and $m_e$ are the effective density and mass of electrons, respectively.

From classical electromagnetic theory, metallic wires behave collectively as small resonant dipoles when excited with an applied electric field parallel to the wires plane, similar to the electric dipoles response of atomic and molecular systems in natural materials [2]. Although the metallic wire structure, discussed earlier, can tailor effective permittivity response within the radio frequency and microwave regime, the arrangement of metallic wires in a cubic structure is still bulky and may appear undesirable for planar radio frequency and microwave applications.

Recently, Falcone et al. introduced in [16–18] a subwavelength resonant planar particle, known as the complementary split-ring resonator (CSRR) as shown in **Figure 4b**, which is the dual counterpart of the SRR. In other words, by following Babinet's principle [19], the complementary of the planar SRR structure is obtained by replacing the SRR metallic rings with apertures and the apertures (surrounding free-space region of SRR) with metal plates. By etching the SRR rings from the metallic ground screen, complementary SRRs form the basis of realizing compact microstrip-based bandstop filters [16]. Such bandstop behavior is attributed to the existence of negative electrical permittivity response, due to an axial time-varying electric field parallel to the CSRR ring plane. Interestingly, as it is practically straightforward to excite CSRR particle with an axial electric field, CSRR particle is very easy to integrate with other planar microstrip circuits.

With the assumption that the largest dimension in CSRR unit cell is much smaller than the operating wavelength, a quasi-static equivalent circuit model can be considered to estimate the effective permittivity response of such inclusion [18].

Consider an axial external uniform time-varying electric field that is parallel to the CSRR inclusion plane, as shown in **Figure 4b**; the composite structure will react and oppose the applied external electric field by creating internal electric dipole moments that give rise to electrical polarization effect. Following the analytical formulation given in [20] and assuming a homogenized artificial CSRR structure, the overall effective electrical permittivity response of the homogenized artificial CSRR structure can then be written in the form below in terms of basic *RLC* circuit elements:

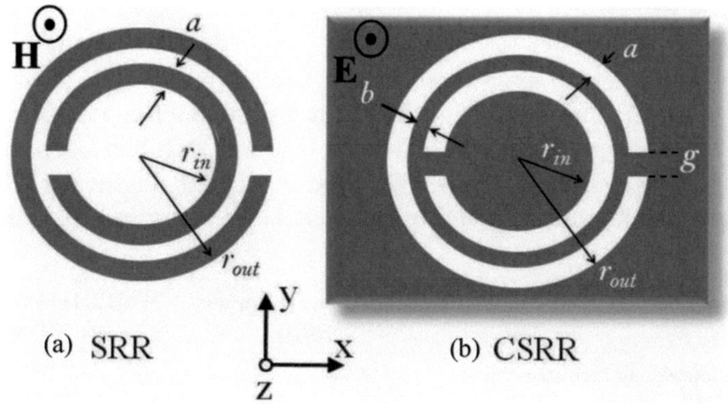


**Figure 4.**
*Two-dimensional view of (a) an artificial magnetic material inclusion, the circular SRR; (b) complementary SRR (CSRR), with dimensions; $r_{out}$ is radius of outer ring, $r_{in}$ is internal ring radius, a represents metallic (aperture) width, b is spacing between rings in SRR (spacing between etched rings in CSRR), and g represents the SRR ring's cut (CSRR etched rings' left metallization). The gray area represents structure metallization.*

$$\varepsilon_{eff} = 1 + \frac{K\, Z_{inc}}{Z_{inc} + \frac{1}{j\omega C_{eff}}} \tag{15}$$

where $K$ is a normalized fractional surface area of CSRR inclusion and $C_{eff}$ is the effective capacitance of a parallel plate capacitor with surface area being the CSRR inclusion surface, while the thickness of the capacitor is the periodicity of the CSRR as a composite infinitely large structure. The term $Z_{inc}$ in Eq. (15) represents the effective impedance of the CSRR inclusion and is given by

$$\mathrm{Z_{inc}} = R_{eff} + j\omega\, L_{eff} \tag{16}$$

where $R_{eff}$ is the effective ohmic losses due to the finite conductivity of the metallic rings around the CSRR slots and is given by the alternating current resistance of the rings. The term $L_{eff}$ in Eq. (16) accounts for the mutual inductive effects between the external and internal strips around the slotted rings. More analytical modeling of the effective electric permittivity and CSRR equivalent circuit parameters can be found in [18, 20]. **Table 1** summarizes the most common artificial magnetic and electric inclusions (i.e., SRR and its dual, CSRR) in order to realize single-μNG and -εNG media, respectively, and their equivalent basic circuit representations (**Figures 5–7**).

### 2.3 Realization of artificial double negative (DNG) medium

In 1968, Veselago investigated theoretically that electromagnetic waves can propagate in a medium, where both permittivity and permeability are negative [4], which was then termed as metamaterials, DNG media, or LHM materials. The realization of artificial DNG media evolved after the theoretical studies by Pendry in [2, 3] in order to synthesize AMMs and artificial ENG media at low frequencies, and afterward practical realizations and demonstrations of DNG media by Smith and his team in [21, 22], which was achieved by producing a large bulk structure composed of repeated patterns of artificial MNG, in the form of SRRs, and ENG media, in the form of planar metallic strips.

Several interesting properties that are not yet in nature can be achieved with DNG media, including but not limited to backward wave propagation and negatively refracted waves and perfect lens [23].

### 2.4 Potential applications of metamaterials

**Table 2** summarizes possible families of artificial metamaterials based on the signs of their permittivity and permeability, assuming such materials are passive, along with a list of suggested engineering applications that had been demonstrated in literature. Metamaterials have been adopted in various engineering fields in wide scope, leading to many existing findings, and are still being explored. It is believed

| Parameter(s) | Artificial magnetic materials (MNG media) | Artificial electric materials (ENG media) |
|---|---|---|
| Commonly adopted unit inclusion | SRR | CSRR |
| Geometry | See **Figure 5(a)** | See **Figure 5(b)** |
| Equivalent circuit topology | See **Figure 6(a)** | See **Figure 6(b)** |
| Typical response to external electromagnetic field | See **Figure 7(a)** | See **Figure 7(b)** |

**Table 1.**
*Commonly adopted MNG and ENG metamaterial building blocks and their equivalent circuit topologies.*

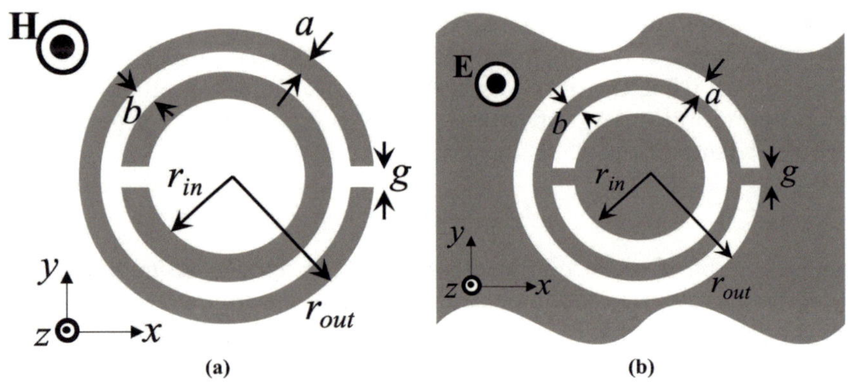


**Figure 5.**
*(a) SRR geometry and (b) CSRR geometry.*

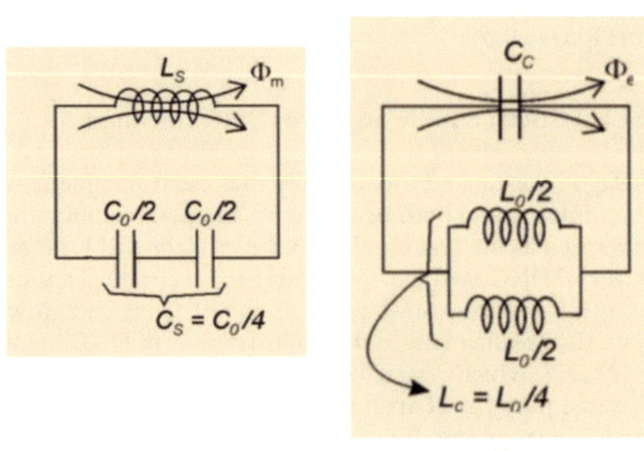


**Figure 6.**
*(a) SRR simplified equivalent circuit model and (b) CSRR simplified equivalent circuit model.*

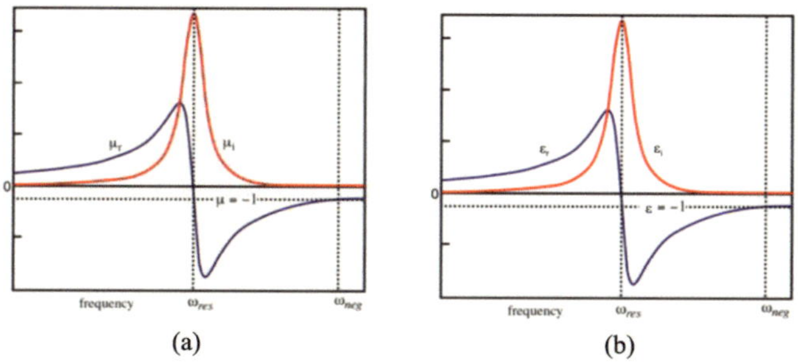

**Figure 7.**
*(a) Expected effective permeability response of SRR and (b) expected effective permittivity response of CSRR.*

| Artificial synthesized metamaterial type | Sign of Re (ε) | Sign of Re (μ) | Sign of Im (ε) | Sign of Im (μ) | Possible engineering applications |
|---|---|---|---|---|---|
| MNG medium | + | - | + | + | • Bandstop/band-pass filters for radio frequency/microwave circuits and systems [24, 25]<br>• Mutual coupling reduction between antennas [26–28]<br>• Electromagnetic shielding of electronic circuits and devices [29]<br>• Gain enhancement of antennas, when used as superstrate [30]<br>• Electromagnetic energy harvesting and absorbers [31, 32] |
| ENG medium | - | + | + | + | • Planar, miniaturized filters for radio frequency/microwave and millimeter wave circuits and systems [16–18]<br>• Mutual coupling reduction between antennas [33]<br>• Electromagnetic shielding of electronic circuits and devices [29, 34, 35] |
| DNG medium | - | - | + | + | • Demonstration of backward wave propagation and negative refraction [21, 22, 36]<br>• Focusing, imaging, and superlensing [23, 37]<br>• Invisibility cloaks [38–40] |

**Table 2.**
*List of general metamaterials families, based on the sign of their constitutive parameters, including possible engineering applications.*

that metamaterials can find lots of applications in other physics and engineering areas, especially in the millimeter and terahertz regimes that might not yet have been reported to date. Note that the number of applications is not limited to those listed in **Table 2**.

## 3. Electromagnetic wave interaction with metamaterials

In this section, the behavior of electromagnetic wave interaction with metamaterials is studied both analytically and numerically. Analytical formulation of electromagnetic field behavior on a one-dimensional artificial lossless and isotropic metamaterial slab due to an external plane wave excitation is presented first. Numerical demonstration of two-dimensional electromagnetic wave interaction with artificial DNG, MNG, and ENG metamaterial slabs is then illustrated and discussed.

## 3.1 Normal plane wave interaction with metamaterials

We consider the problem of a uniform plane wave that is traveling along the $+z$-direction in free space and is incidental normally on an infinitely large metamaterial slab of thickness, $h$. The metamaterial slab is placed between $z = 0$ and $z = h$, as shown in **Figure 8**, and is made infinitely large along $x$ and $y$ directions. For convenience, the incident electric field of the uniform plane wave will be assumed to be in the $+x$ -direction, while its associated magnetic field will be in $+y$ -direction.

Without loss of generality, the phasor notation will be used to express the total electric and magnetic fields in each region of the problem geometry in **Figure 8**. In region 1, the electric and magnetic fields are given as:

$$\mathbf{E}_0 = \left(E_0^+ e^{-jk_0 z} + E_0^- e^{jk_0 z}\right) \mathbf{a}_x \tag{17}$$

$$\mathbf{H}_0 = \left(H_0^+ e^{-jk_0 z} + H_0^- e^{jk_0 z}\right) \mathbf{a}_y \tag{18}$$

where $E_0^+, E_0^-, H_0^+ = E_0^+/\eta_0$, and $H_0^- = -E_0^-/\eta_0$ are the electric and magnetic field amplitudes in the forward and backward directions in region 1 (free space), respectively. The parameters $k_0 = \omega\sqrt{\mu_0\varepsilon_0}$, and $\eta_0 = \sqrt{\mu_0/\varepsilon_0}$ are the wavenumber and wave impedance in free space, respectively.

In the second region, in which the metamaterial slab is located, the phasor form of total electric and magnetic fields is given as

$$\mathbf{E}_1 = \left(E_1^+ e^{-jk_1 z} + E_1^- e^{jk_1 z}\right) \mathbf{a}_x \tag{19}$$

$$\mathbf{H}_1 = \left(H_1^+ e^{-jk_1 z} + H_1^- e^{jk_1 z}\right) \mathbf{a}_y \tag{20}$$

where $E_1^+, E_1^-, H_1^+ = E_1^+/\eta_1$, and $H_1^- = -E_1^-/\eta_1$ are the electric and magnetic field amplitudes in the forward and backward directions in region 2 (metamaterial slab),

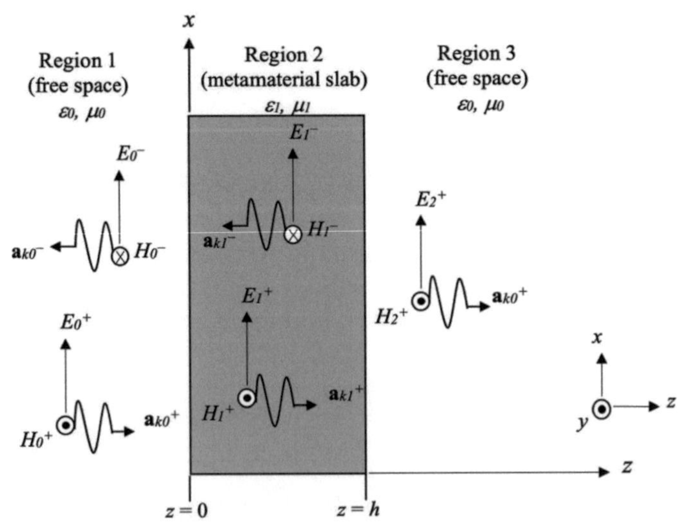


**Figure 8.**
*Normal incidence of a uniform plane wave on an infinitely large metamaterial slab of thickness, h.*

respectively. The parameters $k_1 = \omega\sqrt{\mu_1\varepsilon_1}$ and $\eta_1 = \sqrt{\mu_1/\varepsilon_1}$ are the wavenumber and wave impedance in the metamaterial slab, respectively.

The electric and magnetic fields in region 3, which represents free space, are

$$\mathbf{E}_3 = E_3^+ e^{-jk_0 z}\,\mathbf{a}_x \tag{21}$$

$$\mathbf{H}_3 = H_3^+ e^{-jk_0 z}\,\mathbf{a}_y \tag{22}$$

where $E_3^+$ and $H_3^+ = E_3^+/\eta_0$ are the electric and magnetic field amplitudes in the forward and backward directions in region 3 (free space), respectively.

After applying the boundary conditions at the metamaterial slab walls, $z = 0$ and $z = h$, the following four relations summarizing the relationship between the reflected/transmitted electric field components, $E_0^-$, $E_1^+$, $E_1^-$, and $E_2^+$, respectively, and the incident electric field amplitude, $E_0^+$, are obtained:

$$E_0^- = \frac{j\,(\eta_1{}^2 - \eta_0{}^2)\,\sin(k_1 h)}{2\eta_0\eta_1\cos(k_1 h) + j(\eta_1{}^2 + \eta_0{}^2)\,\sin(k_1 h)}\,E_0^+ \tag{23}$$

$$E_1^+ = \frac{\eta_1(\eta_1 + \eta_0)\,e^{-jk_1 h}}{2\eta_0\eta_1\cos(k_1 h) + j(\eta_1{}^2 + \eta_0{}^2)\,\sin(k_1 h)}\,E_0^+ \tag{24}$$

$$E_1^- = \frac{\eta_1(\eta_0 - \eta_1)\,e^{jk_1 h}}{2\eta_0\eta_1\cos(k_1 h) + j(\eta_1{}^2 + \eta_0{}^2)\,\sin(k_1 h)}\,E_0^+ \tag{25}$$

$$E_2^+ = \frac{2\eta_0\eta_1\,e^{-jk_0 h}}{2\eta_0\eta_1\cos(k_1 h) + j(\eta_1{}^2 + \eta_0{}^2)\,\sin(k_1 h)}\,E_0^+ \tag{26}$$

To simplify Eqs. (23)–(26) further, we define a normalized wave impedance in the metamaterial slab, as $\eta_m = \eta_1/\eta_0 = \sqrt{\mu_r/\varepsilon_r}$ and Eqs. (23)–(26) are read as

$$E_0^- = \frac{j\,(\eta_m{}^2 - 1)\,\sin(k_1 h)}{2\eta_0\eta_1\cos(k_1 h) + j(\eta_m{}^2 + 1)\,\sin(k_1 h)}\,E_0^+ \tag{27}$$

$$E_1^+ = \frac{(\eta_m{}^2 + \eta_m)\,e^{-jk_1 h}}{2\eta_m\cos(k_1 h) + j(\eta_m{}^2 + 1)\,\sin(k_1 h)}\,E_0^+ \tag{28}$$

$$E_1^- = \frac{(\eta_m - \eta_m^2)\,e^{jk_1 h}}{2\eta_m\cos(k_1 h) + j(\eta_m{}^2 + 1)\,\sin(k_1 h)}\,E_0^+ \tag{29}$$

$$E_2^+ = \frac{2\eta_m\,e^{-jk_0 h}}{2\eta_m\cos(k_1 h) + j(\eta_m{}^2 + 1)\,\sin(k_1 h)}\,E_0^+ \tag{30}$$

where $\eta_m$ denotes the normalized wave impedance in the metamaterial slab.

Eqs. (28)–(29) can be used to compute the fractional electric field components in the forward and backward directions, i.e., along the three media. In this analytical study, the electric field strength inside an artificial lossless metamaterial slab is computed using Eq. (19), at a frequency of 10 GHz. We consider the case of a very thin subwavelength homogeneous DNG metamaterial slab, where real parts of permittivity and permeability are both negative at 10 GHz. One possible implementation of such artificial DNG medium can be realized with sufficient number of repeated patterns of composite AMM (SRRs) along with periodic arrangement of metallic rods or planar metallic strips, as demonstrated in [21, 22]. Since evanescent (non-propagating) waves are expected to exist inside single-negative media with exponentially decaying electric field, such cases are not considered here.

**Figure 9** shows the analytically computed electric field strength as a function of the DNG metamaterial slab of thickness, $h$. In this study, the overall DNG slab thickness is considered as $h = 10$ mm. The effect of increasing the effective magnetic permeability of the DNG slab from $-1$ to $-10$ is also presented, as shown in **Figure 9**, where higher mismatch along the interface of DPS-DNG is observed as the effective permeability is increased. The case of matched DNG constitutive parameters with those of DPS (air medium) shows zero reflection from such an interface, as expected. For convenience, the 2D structure is illuminated with an $x$-polarized normal incident plane wave that originates from z = 0 plane. The plane wave has an electric field amplitude peak of 1.5 kV/m and phase of 0°.

For validation purposes, this problem of interest was numerically modeled and simulated using ANSYS HFSS simulator [41]. **Figure 10** presents the developed structure to study the problem of normal plane wave incidence on a one-dimensional DNG metamaterial slab. One possible excitation of such plane wave can be numerically realized using a set of periodic boundary conditions; in other words, using perfect electric conductor (PEC) and perfect magnetic conductor (PMC) symmetry planes along four sides of the geometry, $x$-axis and $y$-axis walls, respectively, ensure plane wave excitation along with proper excitation of the DNG metamaterial slab (see **Figure 10**). Good agreement can be seen between analytically and numerically computed electric field strength inside the DNG metamaterial slab, as shown in **Figures 9** and **11**.

### 3.2 Numerical demonstration of electromagnetic wave interaction with artificial DNG, MNG, and ENG metamaterial slabs

In this section, numerical demonstration of electromagnetic cylindrical wave interaction with various artificial isotropic and lossy metamaterial slabs is illustrated. **Figure 12** depicts the numerical full-wave simulation model, where an artificial lossy DNG slab was placed between two natural, lossy DPS slabs. In this numerical study, DPS, DNG, and MNG media are all considered as lossy and isotropic, where for the case of DNG slab, effective constitutive parameters are $\varepsilon_r = -1$ and $\mu_r = -2.2$, with dielectric and magnetic losses of 0.002, while the constitutive parameters are $\varepsilon_r = 2.2$ and $\mu_r = -1$ for the MNG medium case, and $\varepsilon_r = -2.2$ and $\mu_r = 1$ for the ENG medium slab case, with similar losses as those

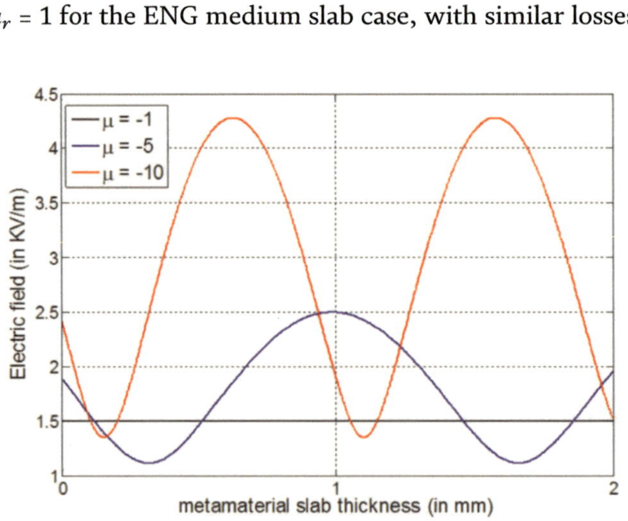


**Figure 9.**
*Analytical computation of electric field strength inside a lossless DNG metamaterial slab. Note that in this study, $\varepsilon_r$ was set as $-1$.*

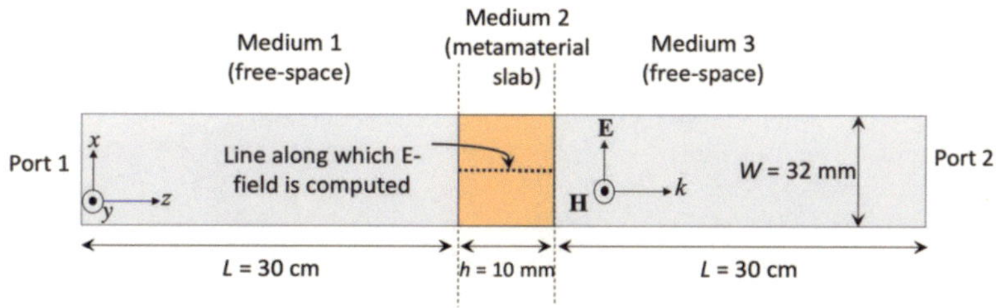


**Figure 10.**
*The numerical full-wave model used to study the normal incidence of plane wave on a one-dimensional DNG lossless metamaterial slab.*

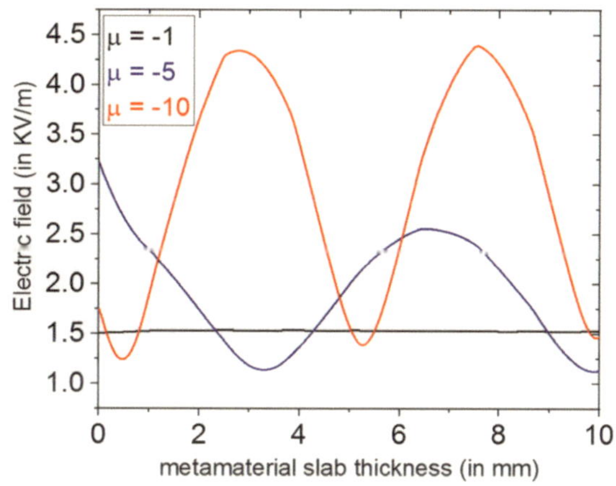


**Figure 11.**
*Numerical computation of electric field strength inside a lossless DNG metamaterial slab. Note that in this study, $\varepsilon_r$ was set as -1.*

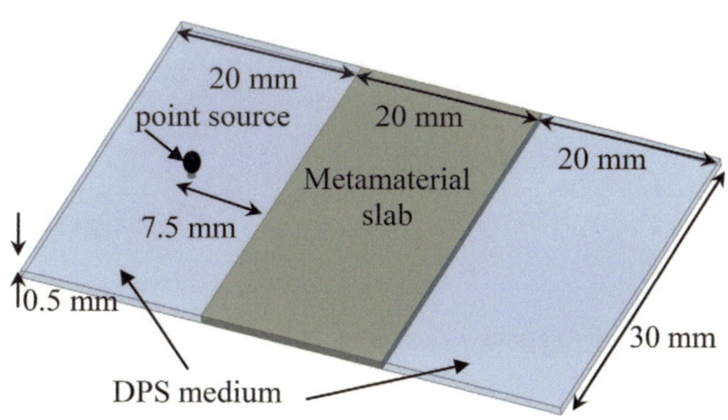


**Figure 12.**
*Numerical model used to study the electromagnetic wave interaction with DNG medium.*

considered in the DNG slab. The constitutive parameters of the DPS medium are $\varepsilon_r = 2.2$ and $\mu_r = 1$, with dielectric and magnetic losses of 0.002. Cylindrical waves were excited from a point source that was placed 7.5 mm away from all the aforementioned slabs. For the electromagnetic wave interaction with MNG medium, the DNG medium in **Figure 12** is replaced with MNG medium. The same is also applied to ENG medium. This numerical demonstration was carried out using the numerical full-wave simulator of ANSYS HFSS.

**Figure 13** presents the electric field intensity distribution for the DNG, MNG, and ENG media that were captured at a phase of 0 degree and compared against the

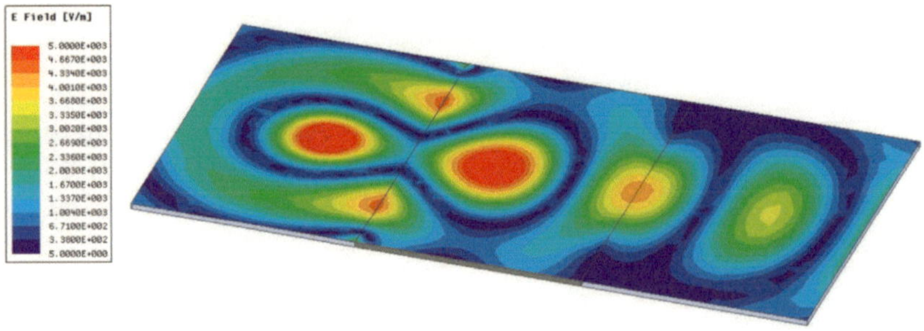


**(a)**

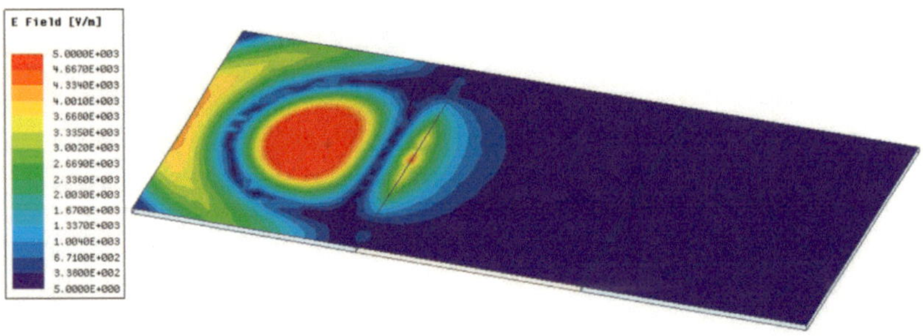


**(b)**

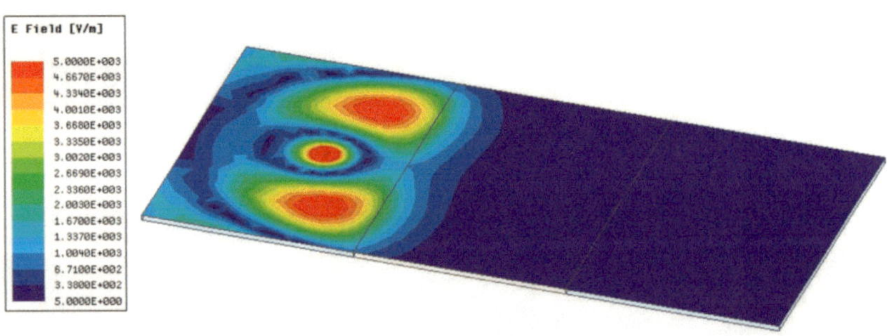


**(c)**

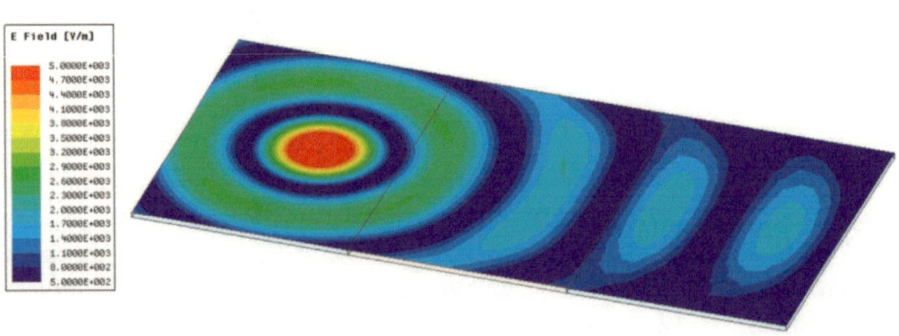


**(d)**

**Figure 13.**
*Numerically computed electric field intensity distribution for the studied lossy isotropic homogenized slabs of (a) DNG medium, (b) MNG medium, (c) ENG medium, and (d) reference DPS medium.*

reference case of lossy DPS slab. In **Figure 13(a)**, which represents the DNG case, several interesting features can be observed, including negatively refracted waves inside the DNG slab and backward wave propagation (see animated figure, **Figure 14(a)** as compared against the DPS case of **Figure 14(b)**), and focusing phenomena of the original electromagnetic cylindrical waves can be seen at the middle of the DNG slab and also at the DPS slab next to the DNG exit face, in which focusing depends on the selection of the refractive indices along with the DNG medium slab thickness.

While electromagnetic wave propagation inside DNG medium is permissible, since both permittivity and permeability are negative and hence result in a positive real wavenumber, evanescent (non-propagating) waves exist in the MNG medium, as shown in **Figure 13(b)**, that only propagate along the interface and decay exponential away from the MNG slab. This is because in the MNG medium, only the permeability is negative, which results in an imaginary negative wavenumber (see Eq. (7)). Similar behavior to the MNG case is also expected for the single-negative ENG medium slab, as shown in **Figure 13(c)**, where evanescent decaying waves are only present in such single-negative medium. For comparison, the case of all DPS media showed normal forward electromagnetic propagation in such lossy media, as shown in **Figure 13(d)**.

A one-dimensional plot of electric field profile inside the aforementioned artificial metamaterials DNG, MNG, and ENG slabs is also presented as shown in **Figure 15** and compared against the normal dielectric DPS medium case. As can be seen from **Figure 15**, propagating electromagnetic field inside the DNG medium

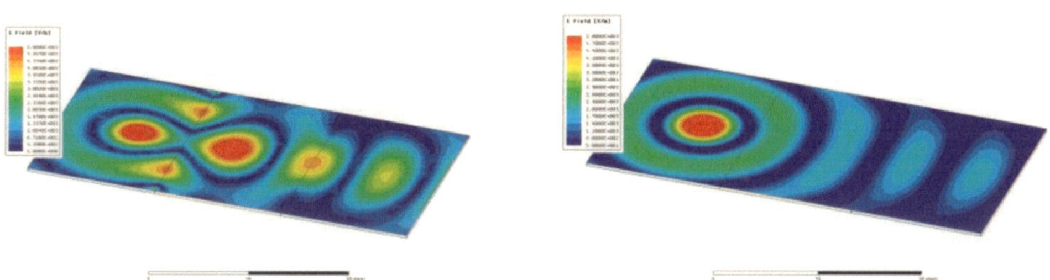

(a) (b)

**Figure 14.**
*Animated snapshots for the electric field distribution for (a) DNG medium and (b) DPS medium cases.*

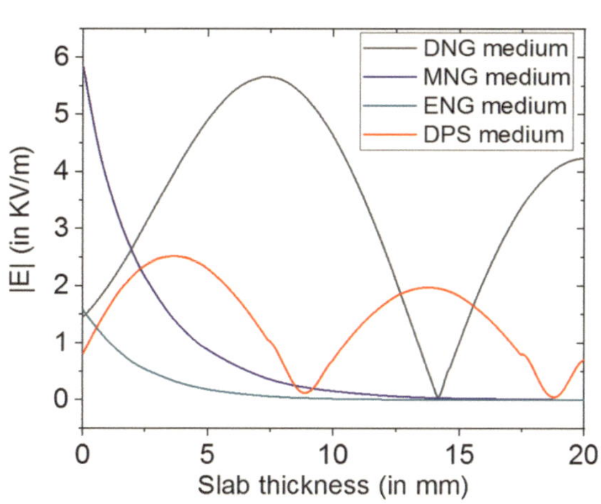


**Figure 15.**
*A one-dimensional plot of electric field strength inside the (a) DNG medium, (b) MNG medium, (c) ENG medium, and (d) DPS medium.*

slab is visible through the recorded electric field profile at a frequency of 10 GHz. Unlike electric field behavior inside the DNG medium, exponentially decaying electric field profile is recorded inside the single-negative MNG and ENG media slabs. Despite the fact that non-propagating (evanescent) waves existed in both MNG and ENG media slabs, the electric field profile is much stronger at the DPS-MNG interface than the strength at the DPS-ENG interface.

## 4. Conclusions

In this chapter, a short review of metamaterials and their realizations based on subwavelength resonant inclusions was presented, along with suggested real-world metamaterial engineering applications that were explored and presented in literature. Over the past 20 years, much interest from researchers and industry was seen to target the use of artificial DNG metamaterials for various engineering applications and physics-based problems. It is important to highlight here that although single-negative media do not permit electromagnetic wave propagation, such artificial media are good candidates for various electromagnetic waves filtering scenarios and harmonic suppression, since they provide high level of electromagnetic wave mitigation and ease of fabrication and integration with radio frequency/ microwave circuits and systems.

Analytical and numerical studies of electromagnetic field behavior and response inside an artificial metamaterial medium were presented. Firstly, the problem of a one-dimensional normal plane wave incidence on an artificial DNG metamaterial slab was analytically formulated, and results were discussed. Furthermore, numerical demonstrations of two-dimensional electromagnetic wave interaction with various lossy metamaterial slabs DNG, MNG, and ENG media were presented and discussed.

## Conflict of interest

The author declares that there is no conflict of interest in the publication of this book chapter.

## Author details

Mohammed M. Bait-Suwailam
Sultan Qaboos University, Muscat, Oman

*Address all correspondence to: msuwailem@squ.edu.om

## References

[1] Jackson JD. Classical Electrodynamics. New York: Wiley; 1998

[2] Pendry JB, Holden AJ, Stewart WJ, Youngs I. Extremely low frequency plasmons in metallic mesostructures. Physical Review Letters. 1996;**76**: 4773-4776

[3] Pendry JB, Holden AJ, Robbins DJ, Stewart WJ. Magnetism from conductors and enhanced nonlinear phenomena. IEEE Transactions on Microwave Theory and Techniques. 1999;**47**:2075-2084

[4] Veselago VG. The electrodynamics of substances with simultaneously negative values of $\varepsilon$ and $\mu$. Soviet Physics Uspekhi. 1968;**10**:509-514

[5] Smith DR, Vier DC, Kroll N, Schultz S. Direct calculation of permeability and permittivity for a left-handed metamaterial. Applied Physics Letters. 2000;**77**:2246-2248

[6] Smith DR, Vier DC, Koschny T, Soukoulis CM. Electromagnetic parameter retrieval from inhomogeneous metamaterials. Physical Review E. 2005;**71**: 036617-036628

[7] Simovski CR, Tretyakov SA. Local constitutive parameters of metamaterials from an effective-medium perspective. Physical Review B. 2007;**75**:195111

[8] Maslovski S, Ikonen P, Kolmakov I, Tretyakov S, Kaunisto M. Artificial magnetic materials based on the new magnetic particle: Metasolenoid. Progress in Electromagnetics Research. 2005;**54**:61-81

[9] Callister WD. Materials Science and Engineering, An Introduction. John Wiley and Sons Inc; 2000

[10] Schelkunoff SA, Friis HT. Antennas: Theory and Practice. New York: Wiley; 1952

[11] Marqués R, Medina F, Rafii-El-Idrissi R. Role of bianisotropy in negative permeability and left-handed metamaterials. Physical Review B. 2002; **65**:144440(1)-144440(6)

[12] Marqués R, Mesa F, Martel J, Medina F. Comparative analysis of edge- and broadside-coupled split ring resonators for metamaterial design-theory and experiments. IEEE Transactions on Antennas and Propagation. 2003;**51**:2572-2581

[13] Baena JD, Marqués R, Medina F, Martel J. Artificial magnetic metamaterial design by using spiral resonators. Physical Review B: Condensed Matter and Materials Physics. 2004;**69**:014402(1)-014402(5)

[14] Yousefi L, Ramahi OM. Artificial magnetic materials using fractal Hilbert curves. IEEE Transactions on Antennas and Propagation. 2010;**58**: 2614-2622

[15] Pines D, Bohm D. A collective description of electron interactions: II. Collective vs individual particle aspects of the interactions. Physical Review. 1952;**85**:338

[16] Falcone F, Lopetegi T, Baena JD, Marqués R, Martín F, Sorolla M. Effective negative-ε stopband microstrip lines based on complementary split ring resonators. IEEE Microwave and Wireless Components Letters. 2004;**14**:280-282

[17] Falcone F, Lopetegi T, Laso MAG, Baena JD, Bonache J, Beruete M, et al. Babinet principle applied to the design of metasurfaces and metamaterials. Physical Review Letters. 2004;**93**: 197401(1)-197401(4)

[18] Baena J, Bonache J, Martín F, Marqués R, Falcone F, Lopetegi T, et al. Equivalent-circuit models for split-ring resonators and complementary split-ring resonators coupled to planar transmission lines. IEEE Transactions on Microwave Theory and Techniques. 2005;**53**:1451-1461

[19] Booker HG. Slot aerials and their relation to complementary wire aerials (Babinet's principle). Journal of the Institute of Electrical Engineers. 1946; **93**(pt. III-A):620-626

[20] Bait-Suwailam MM, Yousefi L, Alavikia B, Ramahi OM. Analytical models for predicting the effective permittivity of complementary metamaterial structures. Microwave and Optical Technology Letters. 2013;**55**: 1565-1569

[21] Smith DR, Padilla WJ, Vier DC, Nemat-Nasser SC, Schultz S. Composite medium with simultaneously negative permeability and permittivity. Physical Review Letters. 2000;**84**:4184-4187

[22] Shelby RA, Smith DR, Schultz S. Experimental verification of a negative index of refraction. Science. 2001;**292**: 77-79

[23] Pendry JB. Negative refraction makes a perfect lens. Physical Review Letters. 2000;**85**:3966-3969

[24] Marqués R, Martín F, Sorolla M. Metamaterials With Negative Parameters: Theory, Design and Microwave Applications. Hoboken, NJ: Wiley; 2008

[25] Gil M, Bonache J, Gil I, García-García J, Martín F. Miniaturization of planar microwave circuits by using resonant-type left handed transmission lines. IET Microwave Antennas and Propagation. 2007;**1**:73-79

[26] Ferrer PJ, Gonzalez-Arbesu JM, Romeu J. Decorrelation of two closely spaced antennas with a metamaterial AMC surface. Microwave and Optical Technology Letters. 2008;**50**:1414-1417

[27] Rajo-Iglesias E, Quevedo-Teruel O, Inclan-Sanchez L. Planar soft surfaces and their application to mutual coupling reduction. IEEE Transactions on Antennas and Propagation. 2009;**57**: 3852-3859

[28] Bait-Suwailam MM, Boybay MS, Ramahi OM. Electromagnetic coupling reduction in high-profile monopole antennas using single-negative magnetic metamaterials for MIMO applications. IEEE Transactions on Antennas and Propagation. 2010;**58**:2894-2902

[29] Celozzi S, Araneo R, Lovat G. Electromagnetic Shielding. Wiley-IEEE Press; 2008. ISBN: 978-0-470-05536-6

[30] Attia H, Yousefi L, Bait-Suwailam MM, Boybay MS, Ramahi OM. Enhanced-gain microstrip antenna using engineered magnetic superstrates. IEEE Antennas and Wireless Propagation Letters. 2009;**8**:1198-1201

[31] Ramahi OM, Almoneef TS, AlShareef M, Boybay MS. Metamaterial particles for electromagnetic energy harvesting. Applied Physics Letters. 2012;**101**:173903

[32] Watts CM, Liu X, Padilla WJ. Metamaterial electromagnetic wave absorbers. Advanced materials. 2012;**24**: 98-120

[33] Bait-Suwailam MM, Siddiqui OF, Ramahi OM. Mutual coupling reduction between microstrip patch antennas using slotted-complementary split-ring resonators. IEEE Antennas and Wireless Propagation Letters. 2010;**9**:876-878

[34] Casey KF. Electromagnetic shielding behavior of wire-mesh screens. IEEE Transactions on Electromagnetic Compatibility. 1988;**30**: 298-306

[35] Bait-Suwailam MM, Ramahi OM. Ultrawideband mitigation of simultaneous switching noise and EMI reduction in high-speed PCBs using complementary split-ring resonators. IEEE Transactions on Electromagnetic Compatibility. 2012;**54**:389-396

[36] Eleftheriades GV, Iyer AK, Kremer PC. Planar negative refractive index media using periodically LC loaded transmission lines. IEEE Transactions on Microwave Theory and Techniques. 2002;**50**:2702-2712

[37] Eleftheriades GV, Siddiqui OF. Negative refraction and focusing in hyperbolic transmission-line periodic grids. IEEE Transactions on Microwave Theory and Techniques. 2005;**53**: 396-403

[38] Schurig D, Mock J, Justice BJ, Cummer SA, Pendry JB, Starr AF, et al. Metamaterial electromagnetic cloak at microwave frequencies. Science. 2006; **314**:977-980

[39] Ergin T, Stenger N, Brenner P, Pendry JB, Wegener M. Three-dimensional invisibility cloak at optical wavelengths. Science. 2010;**328**:337-339

[40] Selvanayagam M, Eleftheriades GV. An active electromagnetic cloak using the equivalence principle. IEEE Antennas and Wireless Propagation Letters. 2012;**11**:1226-1229

[41] ANSYS® Electromagnetics Suite, 17.2 [Online], http://www.ansys.com 熱

Chapter 8

# Metamaterial: Smart Magnetic Material for Microwave Absorbing Material

*Wisnu Ari Adi, Yunasfi Yunasfi, Mashadi Mashadi, Didin Sahidin Winatapura, Ade Mulyawan, Yosef Sarwanto, Yohanes Edi Gunanto and Yana Taryana*

## Abstract

Metamaterial is an artificial, advanced material that has properties such as electromagnetic waves (EM), namely isotropic materials with permittivity and permeability in a single phase at a certain frequency. Smart magnetics is one of the metamaterials that is a modified magnetic material that has a single-phase permeability and permittivity as a function of frequency depending on the type of magnetic material used. Smart magnetics in this study include perovskite, ferrite, hexagonal ferrite, and composite systems. Research that has been carried out on perovskite, ferrite, hexagonal ferrite and composite smart magnetic system materials are $La_{0.8}Ba_{0.2}Fe_xMn_{½(1-x)}Ti_{½(1-x)}O_3$, $Ni_xFe_{3-x}O_4$, and $Ba_{(1-x)}Sr_xFe_2O_4$, $Ba_{0.6}Sr_{0.4}Fe_{12-z}Mn_zO_{19}$ and composite silicon rubber—iron oxide. The four smart magnetic material systems have an average microwave absorption in the X-band frequency range. Very varied reflection loss characteristics depend on the smart magnetic material system formed. It was concluded that smart magnetic material is a microwave absorbent that has reflection loss values in the X-band frequency range. Smart magnetic material is certainly not able to absorb microwaves on all band frequencies because each smart magnetic material has different resonance characteristics, so the maximum effort that can be done is to find the right composition of smart magnetic material which is expected to have the maximum wave absorption capability.

**Keywords:** metamaterial, smart magnetic, microwave, absorbing, perovskite, ferrite, hexagonal ferrite, composite

## 1. Introduction

The rapid progress in communication technology in recent years has been noted by scientists and engineers working in this field. These technological advances have motivated people to utilize them with the purpose to improve their quality of life. A good level of life always strives for ease of communication, among others, by the presence of cellphone abbreviated as hand phone, a type of wireless telephone that is easy to carry everywhere and practical because of its small size so that it is easily inserted into a pocket. A cellular telephone or hand phone is a device that can make and receive phone calls transmitted via electromagnetic waves and can be used



IntechOpen

around a large geographical area. Because communication using this cellphone uses electromagnetic waves in the microwave frequency range, the microwave radiating out of the mobile emitter will theoretically affect the human body, especially the head around the ear as shown in **Figure 1**.

Radiation emitted can also affect the function of enzymes and proteins, which is a change in albumin protein that functions in supplying blood flow to the brain. For this reason, we need a microwave absorbent material that can reduce and even eliminate the effect of microwave radiation on human health [1, 2].

In the electronics field, microwave absorbers are used to reduce the presence of electromagnetic wave interference (EMI) [3, 4]. In general, electronic components that work at high frequencies often experience problems such as frequency signal leakage. EMI will not be present if the electronic device is in an open condition or is not in a closed medium. However, signals travelling in a closed medium will be reflected back to the device. This will cause the energy to increase in phase at certain frequencies due to the appearance of EMI emitted in the form of noise, which then interferes with the performance of these electronic devices. But after the closed media are protected by microwave absorbers, the effect of EMI can be avoided. An illustration of this phenomenon is shown in **Figure 2**.

In the field of defense (military) [5, 6], this microwave absorber is used for coating or painting on defense equipment and facilities such as stealth aircraft, warships (war ship), and for army clothing, especially troops in the guard front. as shown in **Figure 3**.

In a radar system, microwaves are transmitted continuously in all directions by the transmitter. If there is an object affected by this wave, the signal will be reflected by the object and received back by the recipient. This reflection signal will provide information that there is a close object that will be displayed by the radar screen. Radar (radio detection and ranging) is a microwave system that is useful for detecting and measuring distances and making maps of an object. The radar waves emitted are able to detect the presence of an object. The radar concept is measuring the distance from the sensor to the target. The measure of distance is obtained by measuring the time needed by the radar wave during its propagation from the sensor to the target and back to the sensor again. The measured distance based on the time needed by the electromagnetic waves emanating from the target is then reflected back to the radar sensor. The target is able to reflect electromagnetic

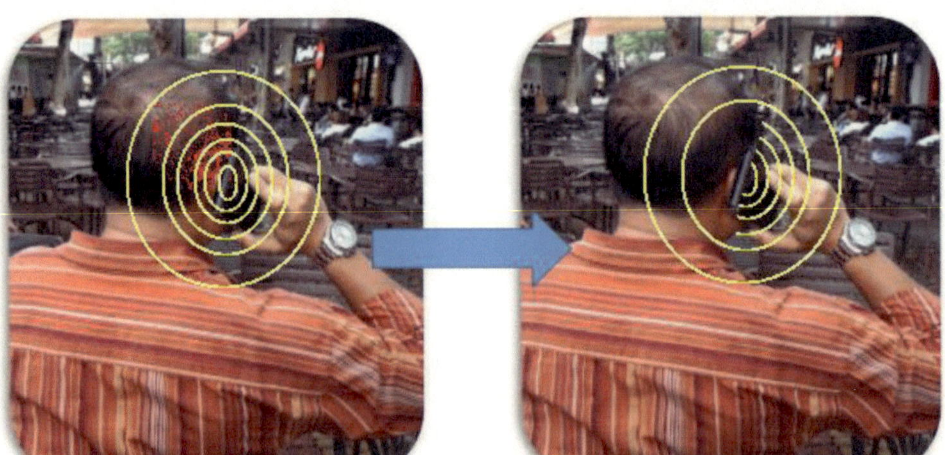

**Figure 1.**
*Use of microwave absorbers for shielding radiation.*

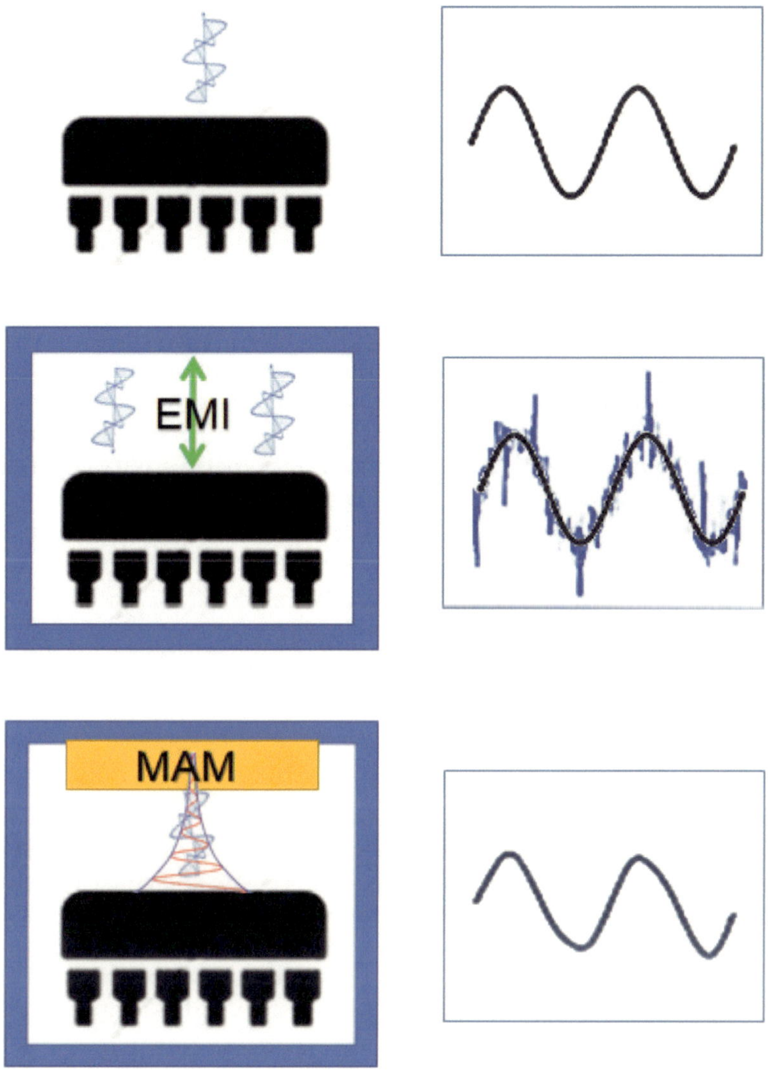


**Figure 2.**
*Use of microwave absorbers to reduce EMI.*

**Figure 3.**
*Use of microwave absorbers in the defense sector.*

waves, so that the radar is able to detect the existence of these objects. However, the case is different if the object cannot reflect radar waves, so that the radar is not able to detect the existence of the object. This phenomenon is then developed for certain interests related to the defense system.

Because of the vast utilization of these microwave absorbent materials, it is generally accepted, recognizing that microwave absorbent material is a material that can weaken the energy of electromagnetic waves. These microwave absorbent materials can externally reduce or even eliminate reflections or transmissions from certain objects and can be used internally to reduce oscillations caused by resonance cavities. Besides that, this microwave absorbent can be used to create a reflection free space or anechoic space.

Metamaterial is one of the solutions for the development of microwave absorbent materials. Metamaterials that are developing rapidly are smart magnetic-based materials. Smart magnetics are modified magnetic materials so these materials have a single-phase frequency-dependent permeability and permittivity depending on the type of magnetic material used. Smart magnetic is an advanced magnetic material in the future in the form of new inorganic crystalline materials with permeability and permittivity made from interpenetrating lattices with magnetic field and electric field responses.

## 2. Conceptual

### 2.1 Electromagnetic wave

Electromagnetic waves are a form of energy emitted and absorbed by charged particles, which shows wavelike behavior because it travels through space [7]. Electromagnetic waves are transverse waves that oscillate and consist of electric field and magnetic field vector components as shown in **Figure 4**.

Electromagnetic energy propagates in waves with several parameters that can be measured, namely, wavelength, frequency, amplitude (amplitude), and speed.

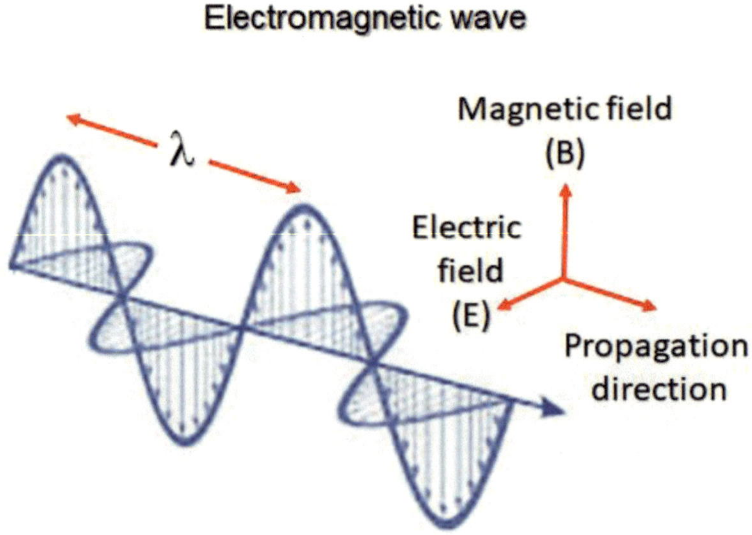


**Figure 4.**
*Schematic propagation of electromagnetic waves.*

Amplitude is the wave height, while the wavelength is the distance between two peaks. Frequency is the number of waves that pass through a point in a unit of time. The frequency depends on the speed of the wave climbed. Because the speed of electromagnetic energy is constant (the speed of light), the wavelength and frequency are inversely proportional. The longer the wave, the lower the frequency, and the shorter the wave, the higher the frequency.

The general characteristics of electromagnetic waves are that changes in the electric and magnetic fields occur at the same time, so that both fields have maximum and minimum values at the same time and at the same place. The direction of the electric field and magnetic field is perpendicular to each other, and both are perpendicular to the direction of wave propagation, electromagnetic waves are transverse waves, and electromagnetic waves experience events of reflection, refraction, interference, polarization, and diffraction. Fast propagation of electromagnetic waves depends only on the electrical and magnetic properties of the medium that it passes through.

### 2.2 Microwave

The arrangement of all forms of electromagnetic waves based on their wavelengths and frequencies covers a very low range of energy to very high energy called the electromagnetic wave spectrum, as shown in **Figure 5**.

Microwaves are electromagnetic waves which have a frequency range of about 0.3–300 GHz with wavelengths of around 1–1 mm. The microwave frequency range consists of several bandwidths, namely, L band to D band [8].

### 2.3 Absorption mechanism

Coherent and polarized microwaves obey the optical law; this wave can be reflected, transmitted, and absorbed, depending on the type of material it passes. In general, the use of microwaves is based on the phenomenon of reflection and transmission only (**Figure 6**).

But in the last few decades, the phenomenon of microwave absorption has also become very popular as the core concept in the development of rapidly advancing electronic and telecommunications technology as shown in core (**Figure 7**) [9–15]. The main requirement that is needed as microwave absorbing material is that this material has a value of permeability (magnetic loss properties) and permittivity (dielectric loss properties) material [16].

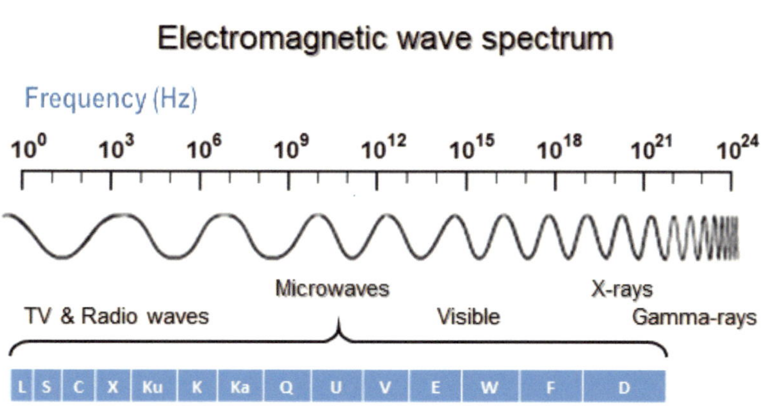


**Figure 5.**
*Spectrum of electromagnetic waves and microwaves.*

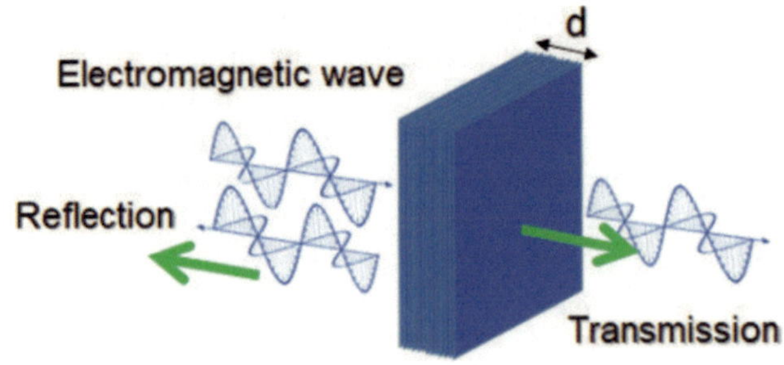


**Figure 6.**
*Rules of optical law in microwaves.*

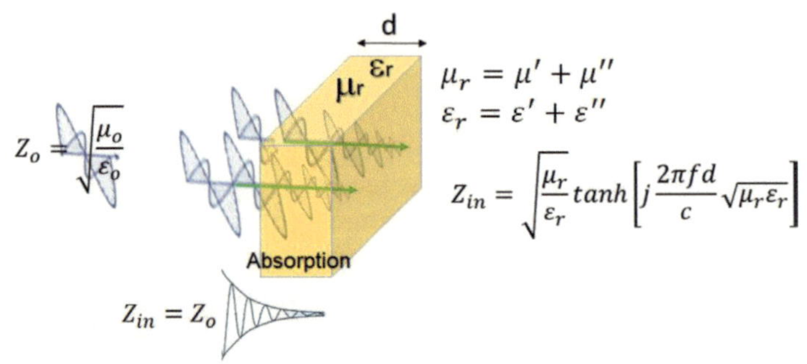


**Figure 7.**
*Microwave absorption mechanism for materials.*

In general the electrical and magnetic properties of a microwave absorbent material are characterized by complex permittivity and complex permeability, as shown by the following equations [17–19]:

$$\varepsilon_r = \varepsilon' + \varepsilon'' \tag{1}$$

$$\mu_r = \mu' + \mu'' \tag{2}$$

The real part of permittivity ($\varepsilon'$) states the measure of the amount of energy from the external electric field stored in the material, while the imaginary part ($\varepsilon''$) states the measure of energy lost due to the external electric field. If the imaginary part is zero then the material is a lossless material and is called a loss factor. The same for permeability, the real part ($\mu'$) expresses a measure of the amount of energy from the external magnetic field stored in the material, while the imaginary part ($\mu''$) shows the amount of energy dissipated due to the magnetic field.

Permittivity is present from material dielectric polarization. The quantity $\varepsilon'$ can also be referred to as the dielectric constant of a material. The quantity $\varepsilon''$ is a measure of attenuation from the electric field caused by material. Loss of tangent permittivity of a material is defined as follows:

$$Tan\ \delta_\varepsilon = \frac{\varepsilon''}{\varepsilon'} \tag{3}$$

The greater the loss tangent of a material, the greater the attenuation when the wave moves through the material. The same applies to magnetic fields, namely:

$$Tan\ \delta_\mu = \frac{\mu''}{\mu'} \tag{4}$$

Both components contribute to the compression of wavelengths in the material. Because electromagnetic waves (EM) are a combination of two waves between electric and magnetic waves, loss of both magnetic and electric fields will weaken the energy in waves. In most dampers, both permittivity and permeability are functions of frequency and can vary significantly even in small frequency ranges. If permittivity and complex permeability are known in a certain frequency range, the material effect on the wave will be known.

It is well known that dielectric and magnetic parameters include electric field vectors $\vec{E}$, magnetic fields $\vec{H}$, induction fields $\vec{B}$, displacement $\vec{D}$, polarization $\vec{P}$, and magnetization $\vec{M}$. The interaction of electric fields in materials follows a pattern similar to magnetic interactions in materials. One of the requirements that must be met for practical application as an absorbent of electromagnetic waves is that this material must have the highest permeability and permittivity values with high magnetic saturation. The SI unit of permittivity and permeability respective are farad per meter and henry per meter. In terms of absorption of EM wave energy, the overall interaction can be represented by the dielectric and magnetic impedance matching of the material (Zin) equal to the air impedance (Zo) as a frequency function.

$$Z_{in} = \sqrt{\frac{\mu_r}{\varepsilon_r}}\, tanh\left[j\frac{2\pi fd}{c}\sqrt{\mu_r \varepsilon_r}\right] \tag{5}$$

where $Z_{in}$ is the impedance of material, ($\mu_r$) and ($\varepsilon_r$) are the complex relative permeability and permittivity of the material, d is the absorber thickness, and c and f are the velocity of light and frequency of microwave in free space, respectively.

$$RL = -20\ log\left|\frac{Z_{in} - Z_o}{Z_{in} + Z_o}\right| \tag{6}$$

This impedance adjustment is important in the microwave frequency range. A transmission line that is given the same load as the characteristic impedance has a standing wave ratio (SWR) equal to one and transmits a certain amount of power without wave reflection. Also the absorption efficiency is optimum if there is no reflected power. Matching means giving the same impedance as the characteristic impedance of electromagnetic waves. Measured parameters are reflection loss (RL), if there is a matching impedance $Z_{in} = Z_o$, meaning that RL will be infinite or all waves have been absorbed perfectly.

## 3. Microwave absorbing material

This chapter focuses specifically on microwave absorbent materials from smart magnet materials which have been thoroughly studied by the authors, which include perovskite, ferrite, hexagonal ferrite, and composite systems. The results have all been reported and published in several globally indexed journals.

### 3.1 Perovskite system

Perovskite systems have an empirical formula $ABO_3$. In this research, the authors have focused on the $LaMnO_3$-based system. An $LaMnO_3$ is a magnetic material that has high permittivity but low permeability because it is paramagnetic at room temperature [20]. In a previous study [21], after $LaMnO_3$ was substituted with barium atoms forming the compound $La_{0.8}Ba_{0.2}MnO_3$, this material was ferromagnetic where the permeability of the material increased. However, the results of testing microwave absorption are still relatively low, only in the range of ~6.5 to ~3 dB at a frequency of 14.2 GHz. In this chapter book, we will also present the results of advanced material engineering based on the results obtained previously, namely, manganite-based materials with a composition of $La_{0.8}Ba_{0.2}Fe_xMn_{½(1-x)}Ti_{½(1-x)}O_3$ (x = 0.1–0.8) [22]. A designed $La_{0.8}Ba_{0.2}Fe_xMn_{½(1-x)}Ti_{½(1-x)}O_3$ composition was prepared using a conventional milling technique. Stoichiometric quantities of analytical grade $BaCO_3$, $Fe_2O_3$, $MnCO_3$, $TiO_2$, and $La_2O_3$ precursors with a purity of greater than 99% were mixed and milled using a planetary ball mill to powder weight ratio of 10:1 for up to 10 h. The quasicrystalline powders were then compacted into pellets and sintered in the electric chamber furnace at 1000°C for 5 h to obtain crystalline materials and confirmed using an X-ray diffractometer (XRD). The results of XRD analysis show that the highest fraction of the $LaMnO_3$ phase was found in the sample with composition $x < 0.3$ to 99%, while the $LaMnO_3$ phase mass fraction decreased for composition $x > 0.3$ as illustrated in **Figure 8**.

The results of magnetic properties analysis were measured using vibrating sample magnetometer (VSM). The results of the VSM analysis show that in all samples, $La_{0.8}Ba_{0.2}Fe_xMn_{½(1-x)}Ti_{½(1-x)}O_3$ (x = 0.1–0.8) contains a $BaFe_{12}O_{19}$-based hard magnetic phase. This magnetic phase increases with increasing composition of x as shown in **Figure 9**.

The characteristic of microwave absorption is measured using vector network analyzer (VNA) in the frequency range 9–15 GHz as illustrated in **Figure 10**. The results of the VNA analysis show that the highest reflection loss is found to be three absorption peaks of ~9, ~8, and ~23.5 dB which is located at 9.9, 12.0, and 14.1 GHz frequency, respectively. Based on the calculation of the reflection value obtained, microwave absorption reaches 95% with a sample thickness of 1.5 mm.

Investigation on this perovskite system has also been carried out by previous researchers. Zhang and Cao [23] succeeded in synthesizing transition metal (TM)-doped $La_{0.7}Sr_{0.3}Mn_{1-x}TM_xO_{3\pm\delta}$ (TM: Fe, Co, or Ni) for microwave absorbing materials. $La_{0.7}Sr_{0.3}Mn_{1-x}TM_xO_{3\pm\delta}$ has shown good properties for

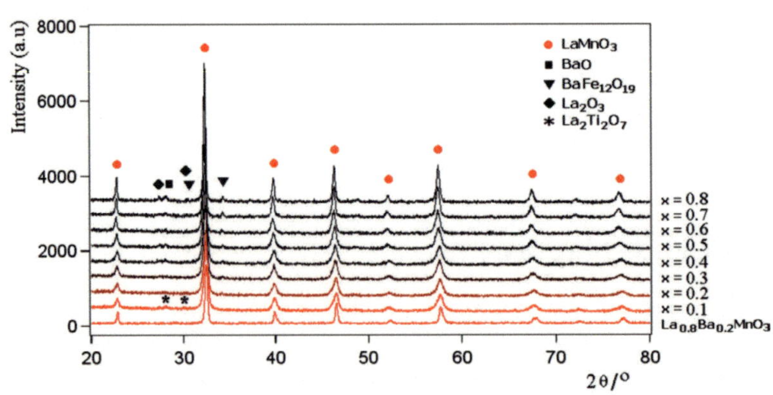


**Figure 8.**
*X-ray diffraction pattern of* $La_{0.8}Ba_{0.2}Fe_xMn_{½(1-x)}Ti_{½(1-x)}O_3$ *(x = 0.1–0.8) [22].*

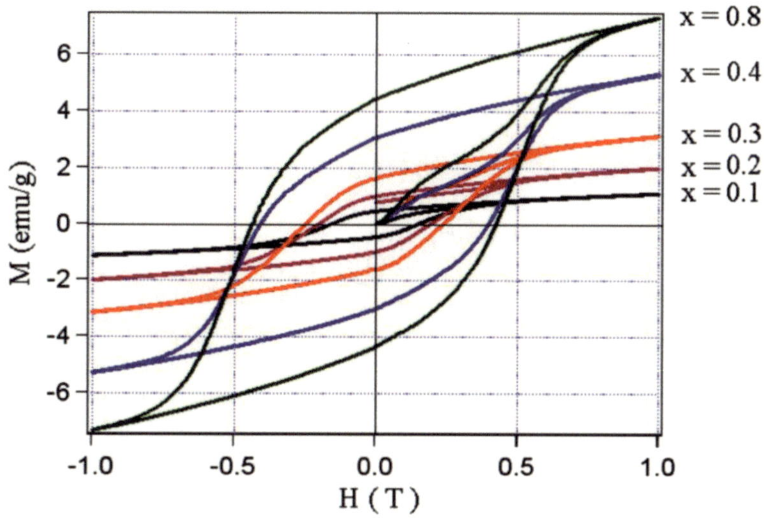


**Figure 9.**
*The hysteresis curve of* $La_{0.8}Ba_{0.2}Fe_xMn_{½(1-x)}Ti_{½(1-x)}O_3$ *(x = 0.1–0.8) [22].*

microwave absorption. The maximum reflection loss was 27.67 dB at a 10.97 GHz frequency, which was obtained from a sample thickness of 2 mm. Zhou et al. [24] reported the successful synthesis of a modified of manganite-based compound $La_{0.8}Sr_{0.2}Mn_{1-y}Fe_yO_3$ ($0 < y < 0.2$). They showed that the absorption bandwidth reached 8.5 GHz above 8 dB and 6.2 GHz above 10 dB; the highest absorption peak reached 34 dB.

### 3.2 Ferrite system

For the ferrite system, we have conducted research on nickel ferrite-based microwave absorbers [25]. A research to study the microwave absorption properties of nickel ferrite in the X-band range has been conducted by using high energy milling technique. The synthesis of nickel ferrite ($Ni_xFe_{3-x}O_4$) was performed using solid-state reaction

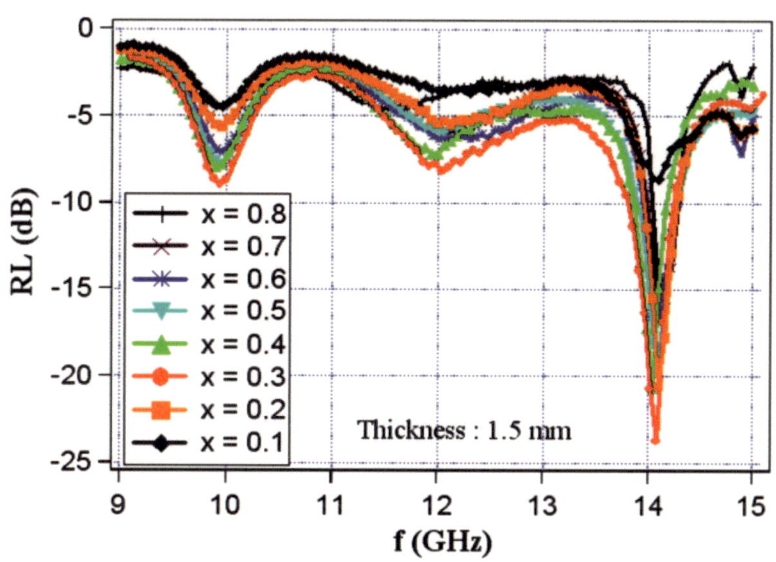


**Figure 10.**
*Reflection loss (RL) of* $La_{0.8}Ba_{0.2}Fe_xMn_{½(1-x)}Ti_{½(1-x)}O_3$ *[22].*

method with the material composition (2x)NiO: (3-x)$Fe_2O_3$ (x = 0.5, 1.0, 1.5 and 2.0) according to the molar ratio. This powder mixture was being milled for 10 hours then sintered at 1000°C temperature for 3 hours. Diffraction patterns of all varied $Ni_xFe_{3-x}O_4$ (x = 0.5, 1.0, 1.5 and 2.0) which have been synthesized by using milling technique are shown in **Figure 11**. It can be noticed that a single phase of all varied $Ni_xFe_{3-x}O_4$ (x = 0.5, 1.0, 1.5 and 2.0), which had spinel structure with lattice parameters a = b = c (space group Fd3m), has successfully formed.

The magnetic properties were measured by using vibrating sample magnetometer (VSM) as shown in **Figure 12**. To study the effects of $Ni^{2+}$ doping on saturation magnetization (Ms), coercivity (Hc), and remanent magnetization (Mr) of $Ni_xFe_{3-x}O_4$ of (x = 0.5–2.0), M-H hysteresis loops were recorded using VSM under the applied magnetic field in the range of −10 up to 10 kOe at room temperature. The VSM result shows that all the samples exhibited a ferromagnetic behavior and fine hysteresis loops with a decrease in magnetization (Ms and Mr) but coercivity (Hc) with increase in $Ni^{2+}$ concentration. Its coercivity value is in the range of 164–217 Oe, and the maximum value is found at x =1.5 composition.

The microwave absorption measurement was carried out by Vector Network Analyzer (VNA). The VNA characterization shows the ability of microwave absorption with a parameter of RL (reflection loss) value. **Figure 13** shows that the highest RL peak reached −28 dB at frequency of 10.98 GHz. It means that the $Ni_{1.5}Fe_{1.5}O_4$ sample can absorb microwave about ~96% at 10.98 GHz.

Other ferrite materials, such as barium mono-ferrite-based microwave absorbers, have also been studied by Ade Mulyawan et al. [26]. Barium mono-ferrite ($BaFe_2O_4$) has a more complex structure that exhibits orthorhombic structure. In this study, barium strontium mono-ferrite has been successfully synthesized using mechanical milling technique. $BaCO_3$, $SrCO_3$, and $Fe_2O_3$ powders are each weighed in accordance with the mole ratio of a total weight of 10 g. The chemical composition for the $Ba_{(1-x)}Sr_xFe_2O_4$ samples was in the range of $0 < x < 0.5$. X-ray diffraction patterns of all varied $Ba_{(1-x)}Sr_xFe_2O_4$ ($0 < x < 0.5$) show a single phase of all composition.

**Figure 14** shows the RL result of $Ba_{(1-x)}Sr_xFe_2O_4$ ($0 < x < 0.5$). The results of VNA analysis show that the highest reflection loss of the $Ba_{(1-x)}Sr_xFe_2O_4$ ($0 < x < 0.5$) was

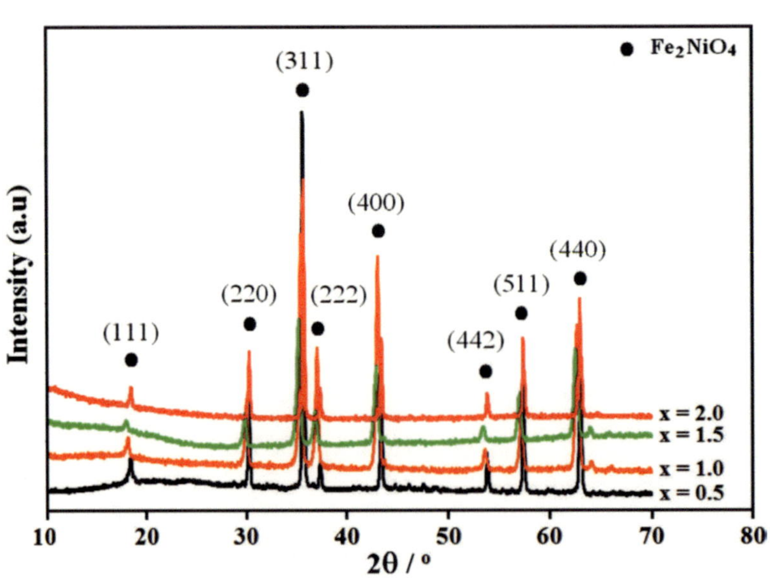


**Figure 11.**
*X-ray diffraction patterns of $Ni_xFe_{3-x}O_4$ of (x = 0.5–2.0) [25].*

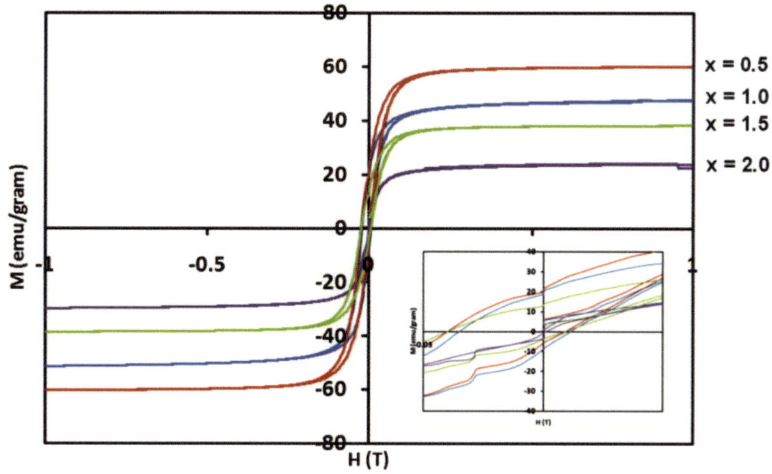


**Figure 12.**
*Hysteresis curve of $Ni_xFe_{3-x}O_4$ ($x$ = 0.5–2.0) [25].*

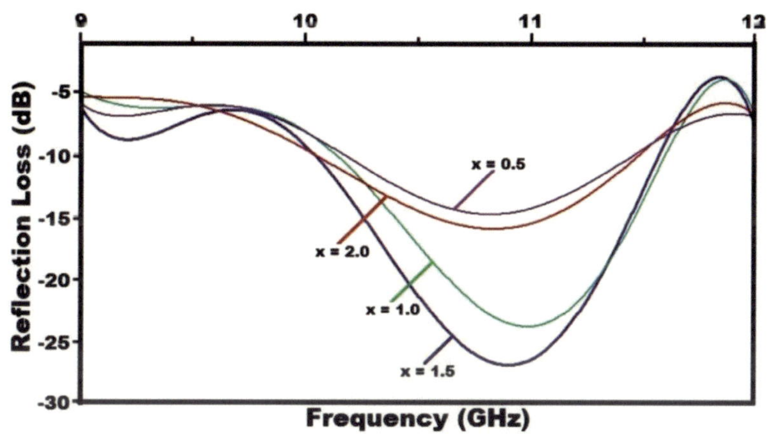


**Figure 13.**
*Reflection loss (RL) curve of $Ni_xFe_{3-x}O_4$ ($x$ = 0.5–2.0) [25].*

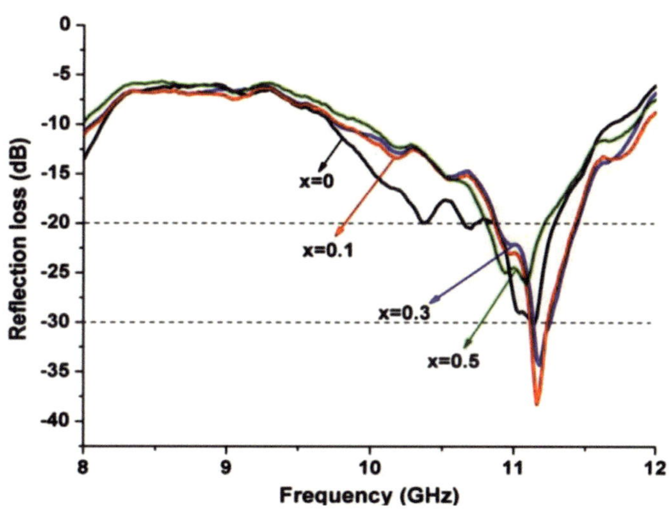


**Figure 14.**
*Reflection loss (RL) curve of $Ba_{(1-x)}Sr_xFe_2O_4$ ($0 < x < 0.5$) [26].*

found in the sample with composition x = 0.1, while the reflection loss of the $Ba_{(1-x)}Sr_xFe_2O_4$ (0 < x < 0.5) decreased for composition x > 0.1. A significant property of microwave absorption has also been displayed for the composition of x = 0.1, in which the value of −38.25 dB (~99.9%) for the reflection loss in the frequency range of 11.2 GHz was achieved.

### 3.3 Hexagonal ferrite system

The M-type hexagonal ferrite system ($BaFe_{12}O_{19}$) is generally an oxide permanent magnet. The hexagonal ferrite is one of the hard magnetic materials that is widely used in many applications. Apart from being applied to electric motors, they can be used for electronic devices because they have good phase stability at high temperatures and very high frequency responses and as switching with narrow field distribution. Material engineering for this application requires material that has magnetic and electrical specifications, and to obtain it, system modification is needed through a substitution process where trivalent iron ions $Fe^{3+}$ will be replaced in part by $M^{2+}$ divalent and $M^{4+}$ tetravalent metal ions; using a material processing route can vary. We have succeeded in modifying this material into a microwave absorbent material. $Ba_{0.6}Sr_{0.4}Fe_{12-z}Mn_zO_{19}$ (z = 0, 1, 2, and 3) was successfully synthesized by solid-state reaction through a mechanical milling method [27]. The raw materials of $MnCO_3$, $BaCO_3$, $Fe_2O_3$, and $SrCO_3$ pro-analytic with purity > 99% were mixed according to stoichiometry composition of $Ba_{0.6}Sr_{0.4}Fe_{12-z}Mn_zO_{19}$ (z = 0, 1, 2, and 3). Based on the results of quantitative analysis using XRD, it shows that the best phase composition was found in the composition z = 1, namely, $Ba_{0.6}Sr_{0.4}Fe_{11}MnO_{19}$ as shown in **Figure 15**. Refinement of X-ray diffraction patterns reveals that $Ba_{0.6}Sr_{0.4}Fe_{11}MnO_{19}$ is a single phase and has a hexagonal structure with space group P63/mmc.

Powder $Ba_{0.6}Sr_{0.4}Fe_{11}MnO_{19}$ has an average particle size of 850 nm. Magnetic properties of $Ba_{0.6}Sr_{0.4}Fe_{11}MnO_{19}$ have a relatively low coercivity field and high remanent magnetization as shown in **Figure 16**.

The results of the microwave absorption test in the sample $Ba_{0.6}Sr_{0.4}Fe_{11}MnO_{19}$ in the frequency range 8–14 GHz show that the absorption peak values were −8 and −10 dB at 8.5 and 12.5 GHz, respectively (**Figure 17**). When compared with the results of the study by Azwar Manaf et al. [28], they have conducted research on the $Ti^{2+}$-$Mn^{4+}$ ions which substituted $BaFe_{12-2x}Ti_xMn_xO_{19}$ samples with x = 0.0–0.8 through a mechanical alloying process and have studied the effect of ion substitution on

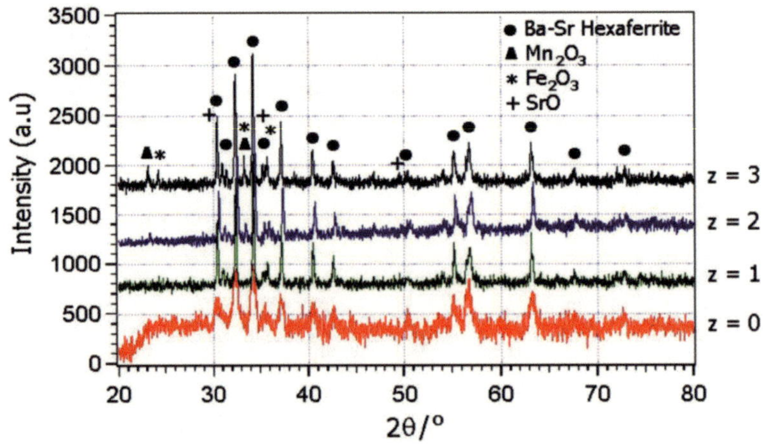


**Figure 15.**
*X-ray diffraction patterns of $Ba_{0.6}Sr_{0.4}Fe_{12-z}Mn_zO_{19}$ (z = 0, 1, 2, and 3) [27].*

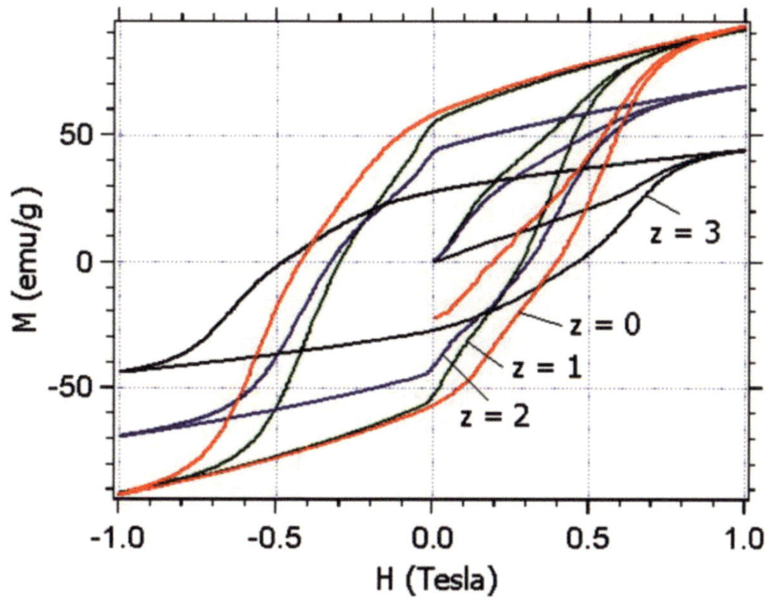


**Figure 16.**
*Hysteresis curve of $Ba_{0.6}Sr_{0.4}Fe_{12-z}Mn_zO_{19}$ (z = 0, 1, 2, and 3) [27].*

microstructure, magnetic, and microwave absorption characteristics. They obtained the results of reflection loss (RL) on series of $Ti^{2+}$-$Mn^{4+}$ ions substituted $BaFe_{12-2x}Ti_xMn_xO_{19}$ samples with x = 0.0–0.8 samples which could be increased from 2.5 dB in composition x = 0 to −22 dB in composition x = 0.6 in the 8–12 GHz frequency range.

The results of this study can be concluded that the modified hexagonal ferrite system is also a good candidate for microwave absorbing material.

### 3.4 Composite system

At present there have been many materials developed from other types of polymer-based composite materials, because application requires that these

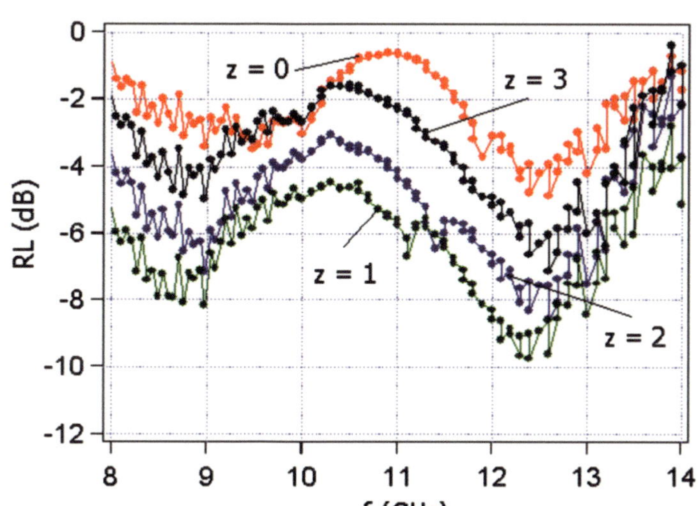


**Figure 17.**
*Reflection loss curve (RL) of $Ba_{0.6}Sr_{0.4}Fe_{12-z}Mn_zO_{19}$ (z = 0, 1, 2, and 3) [27].*

materials should be easy to form and easy to apply to other media such as electronic devices. For composite systems we have conducted research on microwave absorber sheets that are a composite of silicon rubber—iron oxide [29]. In this study, composites from the raw materials of silicone rubber, toluene, and iron oxide magnet powder have been made. The three raw materials are blended in a beaker and stirred for 60 min. The mixture is then printed on a media at 70°C and left for 15 min. The results of elemental analysis using energy-dispersive spectroscopy (EDS) show that the sample contained carbon, oxygen, sulfur, copper, and iron. While the results of the analysis of X-ray diffraction patterns show that the sample was classified as semi-crystalline with a crystallinity of 46%. So the composite consists of an amorphous matrix and a crystalline filler (namely, the phases of $CuFeS_2$, FeS, $FeO_2$, and Fe). Based on the results of functional group analysis using Fourier transformation infrared (FTIR), it shows that the sulfur addition modifies the polymer by forming a cross bond (bridge) between the individual polymer chain and the bond between the magnetic filler and the rubber matrix. **Figure 18** represents the FTIR spectrum between 4000 and 600 $cm^{-1}$ from the composite sample. The peak transmittance of the FTIR spectrum in the composite sample shows the vibrations of O—H, C—H, Si—C, Si—O, and Fe—O bond. Natural rubber consists of suitable polymers of isoprene organic compounds with minor impurities from organic compounds and other water. The transmittance peaks of silicone rubber appear at wave numbers around 3000, 1250, and 1050–750 $cm^{-1}$ which indicate the presence of successive functional groups C—H, Si—C, and Si—O bonds. Peak transmittance oxide iron was also observed at wave numbers around 3700 and 600 $cm^{-1}$, each of which indicated a vibration of H—O and Fe—O bonds. From the results of FTIR analysis, it is suspected that there is a bond between iron oxide as a filler and silicone rubber as a matrix.

The results of magnetic property analysis on absorber sheets were carried out using a vibrating sample magnetometer (VSM) which produced magnetic particle hysteresis curves as shown in **Figure 19**.

The reflection loss value which is a microwave absorbent indicator from the absorber sheet is shown in **Figure 20** where each profile for sheet-1, sheet-2, and sheet-3 has been compared. **Figure 20** shows the relationship between reflection loss (RL) of the absorber sheet and the X-band microwave frequency in the range 10–15 GHz which was measured at a sample thickness of 1.5 mm. There were two observed absorption peaks and high RL in the high frequency range for all samples.

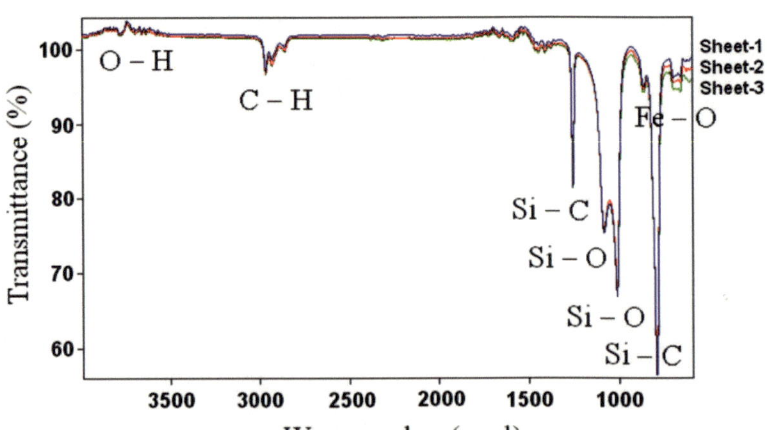


**Figure 18.**
*The transmittance spectrum (FTIR) of the composite sample of silicon rubber—iron oxide [29].*

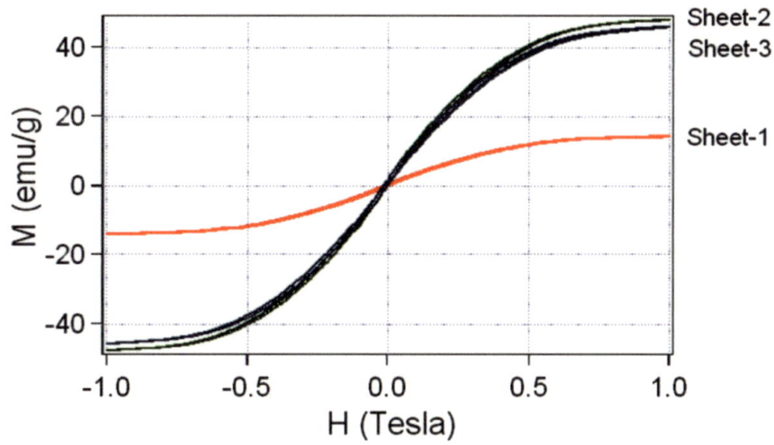


**Figure 19.**
*Hysteresis curve of composite silicon rubber—iron oxide samples [29].*

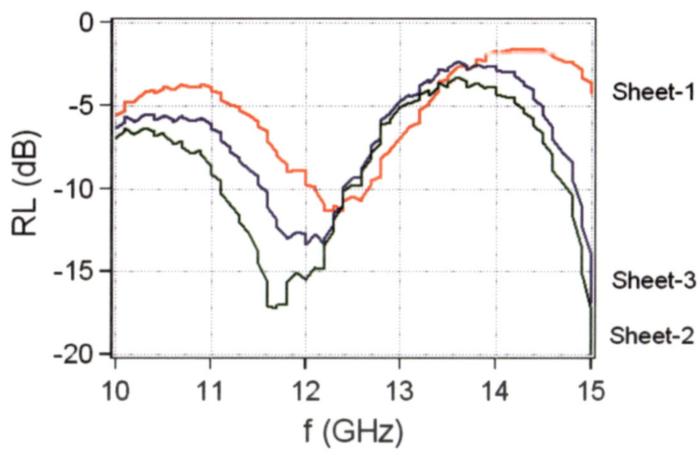


**Figure 20.**
*Reflection loss curve (RL) of composition of silicone rubber—iron oxide [29].*

The RL value increases with the addition of filler composition in the matrix. It seems that the RL value increases with the same sample thickness and the absorption frequency slightly shifts. This can be explained by the effects of electromagnetic properties on attenuation characteristics in each sample. Thus according to the results of this study, it was found that the best composite sample was sheet-3 with an absorption peak value of −15 dB at a frequency of 12 GHz.

Based on the results of this study, we conclude that all microwave absorbent materials can be made in the form of silicon rubber-based composite sheets as a composite matrix.

## 4. Conclusions

Based on the above explanation, wave absorbing materials are composed of materials that have magnetic and electrical properties which are shown by intrinsic parameters in the form of complex permittivity ($\varepsilon_r$) and complex permeability ($\mu_r$) and extrinsic parameters in the form of geometry factors (thickness) of a material [14–16]. If a travelling electromagnetic wave is incident upon and absorbed by a microwave absorbent, spin resonance will occur due to the presence of these material parameters. In smart magnetic materials, the resonance that occurs between electromagnetic waves and material is divided into two mechanisms, namely, the wall resonance domain and spin electron resonance (ferromagnetic resonance). The wall resonance domain is resonance that occurs in magnetic domains caused by induction of electromagnetic waves, while spin electron resonance is resonance that occurs in electrons that are precise in the direction of the internal magnetic field due to the induction of electromagnetic waves. However, it should be noted that smart magnetic material is certainly unable to absorb microwaves on all band frequencies because each smart magnetic material has different resonant characteristics, so the maximum effort that can be done is to find the right composition of smart magnetic material which is expected to have the maximum wave absorption ability.

## Acknowledgements

This work is supported by the program for research and development of smart magnetic material, Center for Science and Technology of Advanced Materials, National Nuclear Energy Agency. The authors are thankful to Prof. Dr. Ridwan and Dr. Abu Khalid Rivai, M.Eng., for their permission to publish this book chapter. Many thanks to Dra. Mujamilah, M.Sc., and Drs. Bambang Sugeng for their kind help to characterize using VSM and XRD, respectively.

## Author details

Wisnu Ari Adi[1*], Yunasfi Yunasfi[1], Mashadi Mashadi[1], Didin Sahidin Winatapura[1], Ade Mulyawan[1], Yosef Sarwanto[1], Yohanes Edi Gunanto[2] and Yana Taryana[3]

1 Center for Science and Technology of Advanced Materials, National Nuclear Energy Agency, Tangerang, Selatan, Indonesia

2 Department of Biology Education, University of Pelita Harapan, Tangerang, Indonesia

3 Center for Research of Electronic and Telecommunication, Indonesian Institute of Sciences, Bandung, Indonesia

*Address all correspondence to: dwisnuaa@batan.go.id

## References

[1] Victorya RM. Effects of handphone's electromagnetic wave exposure on seminiferous tubules. Jurnal Majority. 2015;**4**(3):96-100

[2] Idayati R. Effect of hand phone radiation on the human health. Jurnal Kedokteran Syiah Kuala. 2011;**11**(2):118-123

[3] Eswaraiah V, Sankaranarayanan V, Ramaprabhu S. Functionalized graphene–PVDF foam composites for EMI shielding. Macromolecular Materials and Engineering. 2011;**296**:1-5

[4] Wu Z, Li M. Electromagnetic interference shielding of carbon nanotube macrofilms. Scripta Materialia. 2011;**64**:809-812

[5] Saville P. Review of radar absorbing materials. Defence R&D Canada–Atlantic, Technical Memorandum, DRDC Atlantic TM. 2005;**2005-003**

[6] Vinoy KJ, Jha RM. Radar Absorbing Materials: From theory to Design and Characterization. Boston: Kluwer Academic Publishers; 1996

[7] Kovetz A. Electromagnetic Theory. Oxford: New York; 2000

[8] Chen LF et al. Microwave Electronics: Measurements and Materials Characterization. Chichester: John Wiley and Sons; 2004

[9] Wu C, Chen S, Gu X, Hu R, Zhong S, Tan G, et al. Enhanced and broadband absorber with surface pattern design for XBand. Current Applied Physics. 2018;**18**:55-60

[10] Nam Y-W, Choi J-H, Lee W-J, Kim C-G. Fabrication of a thin and lightweight microwave absorber containing Ni-coated glass fibers by electroless plating. Composites Science and Technology. 2017;**145**:165-172

[11] Salman S, Afghahi S, Jafarian M, Atassi Y. Novel approach for designing a thin and broadband microwave absorber in Ku band based on substituted M-hexaferrites. Journal of Magnetism and Magnetic Materials. 2016;**419**:62-67

[12] Qing Y, Zhou W, Luo F, Zhu D. Thin-thickness FeSiAl/flake graphite-filled $Al_2O_3$ ceramics with enhanced microwave absorption. Ceramics International. 2017;**43**:870-874

[13] Mashadi Y, Adi WA. Structural, magnetic properties and microwave absorbing behavior of iron based carbon nanocomposites. Malaysian Journal of Fundamental and Applied Sciences. 2016;**12**(1):6-11. ISSN: 2289-5981

[14] Adi WA, Sarwanto Y, Taryana Y, Soegijono B. Effects of the geometry factor on the reflection loss characteristics of the modified lanthanum manganite. IOP Conference Series. Journal of Physics: Conference Series. 2018;**1091**:012028. DOI: 10.1088/1742-6596/1091/1/012028

[15] Adi WA, Wardiyati S, Dewi SH. Nanoneedles of lanthanum oxide ($La_2O_3$): A novel functional material for microwave absorber material. IOP Conference Series: Materials Science and Engineering. 2017;**202**:012066. DOI: 10.1088/1757-899X/202/1/012066 012076

[16] Jiu R et al. Complex permittivity, permeability and electromagnetic wave absorption of α-Fe/C(amorphous) and $Fe_2B$/C(amorphous) nanocomposites. Journal of Physics D: Applied Physics. 2004;**37**:2737-2741. DOI: 10.1088/0022-3727/37/19/019

[17] Folgueras LC, Alves MA, Rezende MC. Microwave absorbing paints and sheets based on carbonyl iron and polyaniline: Measurement and simulation of their properties.

Journal of Aerospace Technology and Management. 2010;**2**(1)

[18] Rohde, Schwarz. Measurement of Dielectric Material Properties. Application Center Asia/Pacific. CY Kuek 07; 2006

[19] Hecuadi AY, Adi WA, Purwanto S. A study on the behaviour of M-type strontium hexagonal ferrite based microwave absorbing. In: Proceedings of the 6th National Radar Seminar and ICRAMET 2012. Bali, Indonesia; 2012

[20] Mondal P, Bhattacharya D, Choudhury P. Dielectric anomaly at the orbital order–disorder transition in $LaMnO_3+\delta$. Journal of Physics: Condensed Matter. 2006;**18**(29):6869-6881

[21] Sardjono P, Adi WA. Thermal analysis and magnetic properties of lanthanum barium manganite perovskite. Journal of Advanced Materials Research. 2014;**896**:381-384

[22] Adi WA, Manaf Ridwan A. Absorption characteristics of the electromagnetic wave and magnetic properties of the $La_{0.8}Ba_{0.2}FexMn_{½(1-x)}Ti_{½(1-x)}O_3$ (x = 0.1-0.8) Perovskite system. International Journal of Technology. 2017;**5**:887-897. DOI: 10.14716/ijtech.v8i5.871

[23] Zhang S, Cao Q. Electromagnetic and microwave absorption performance of some transition metal doped $La_{0.7}Sr_{0.3}Mn_{1-x}TM_xO_{3\pm1}$ (TM = Fe, Co or Ni). Materials Science and Engineering B. 2012;**177**:678-684

[24] Zhou KS, Xia H, Huang K-L, Deng L-W, Wang D, Zhou Y-P, et al. The microwave absorption properties of $La_{0.8}Sr_{0.2}Mn_{1-y}Fe_yO_3$ nanocrystalline powders in the frequency range 2-18 GHz. Physica B: Condensed Matter. 2009;**404**:175-179

[25] Yunas Y, Adi WA, Mashadi M, Rahmy PA. Magnetic and microwave absorption of nickel Ferrite ($Ni_xFe_{3-x}O_4$) by HEM technique. Malaysian Journal of Fundamental and Applied Sciences. 2017;**13**(3):2013-2016. DOI: 10.11113/mjfas.v13n3.553

[26] Mulyawan A, Ari W, Yunasfi A. Raman spectroscopy study, magnetic and microwave absorbing properties of modified barium strontium monoferrite $Ba_{(1-x)}Sr_{(x)}Fe_2O_4$. Malaysian Journal of Fundamental and Applied Sciences. 2018;**14**(1):73-77. DOI: 10.11113/mjfas.v14n1.750

[27] Gunanto YE, Jobiliong E, Adi WA. Microwave absorbing properties of $Ba_{0.6}Sr_{0.4}Fe_{12-z}Mn_zO_{19}$ (z = 0-3) materials in X band frequencies. Journal of Mathematical and Fundamental Sciences. 2016;**48**(1):55-65

[28] Manaf A, Elita Hafizah MA, Belyamin B, Nainggolan B, Manawan M. Magnetic and microwave absorption characteristics of $Ti^{2+}$-$Mn^{4+}$ substituted barium hexaferrite. International Journal of Technology. 2017;**8**(3). DOI: 10.14716/ijtech.v8i3.5416

[29] Didin S, Winatapura D, Adi WA. Microwave absorber sheet of the composite silicon rubber–iron oxide. IOP Conference Series. Journal of Physics: Conference Series. 2018;**1091**:012006. DOI: 10.1088/1742-6596/1091/1/012006

Section 4

# The Effect of Electromagnetic Waves

Chapter 9

# Effects of Electromagnetic Field on the Development of Chick Embryo: An In Vivo Study

*Najam Siddiqi and Nasser Al Nazwani*


## Abstract

This study was conducted to explore the effects of electromagnetic waves on a developing chick embryo. The radiofrequency electromagnetic waves (RFW) emitted by different smart phones was measured by using a TriField meter. Chick fertilized eggs were placed in an egg incubator, divided into control and exposed groups. In the exposed group, a mobile phone was placed inside an incubator in call receiving mode, while in the control group, the mobile phone was not used. Studies were conducted at low and high exposure (dose) of RFW. Chick embryos were sacrificed at day 10 and day 15, and embryos were examined for mortality, gross malformation, weight, and length. Histology, electron microscopy, and Hsp 70 of liver were done for the high dose group. No mortality was observed in the low dose group; however, in the high dose group, the mortality was 14%, and deformities of the limbs and skin abnormalities were observed. Weight and length in the exposed groups were significantly lower than the control at higher dose. Histology and ultrastructure of liver revealed fatty infiltration, increase number of mitochondria, deformation, and disappearance of its cristae. Hsp 70 and mRNA levels were elevated in the exposed groups for high dose group.


**Keywords:** electromagnetic waves, mobile phone, chick embryo, liver, fatty change, mitochondria, ultrastructure, mortality, gross morphology, histology

## 1. Introduction

Cellular technology and broadband services are growing rapidly, resulting in a dense suffusion of nonionizing low radio frequency electromagnetic waves (RFW) in the atmosphere. This adds a new dimension to environmental pollution. Radiofrequency electromagnetic radiation is a form of energy used in wireless communication and emitted from mobile sets which can be absorbed into body tissues and converted into heat.

These waves were initially thought to be harmless to the humans; however, now scientific research has revealed that these waves may cause damage to the living cells. Smart phone emits radio waves while in use which include downloading data from the internet [1, 2]. Fetus and children are more radiosensitive than adults due to the presence of embryonic stem cells [3–7]. A child born in this era will start to be exposed to electromagnetic waves exposure as soon as he is born and will remain in this environment until he dies. Divan et al. reported behavioral problems in children who were exposed to prenatal and postnatal cell phone [8, 9]. This potential hazard is more serious in developing countries, where the reported prevalence of mobile phone use by adolescents is more than 90%. Oman is ranked 10th in the world after Saudi Arabia, Russia, Kuwait, and Panama [10, 11]. Excessive use of mobile phones and base stations in the vicinity of residential areas are linked to symptoms such as headaches, sleep disturbances, lack of concentration, dizziness, memory loss, and increased risk of cancer were first reported as "Microwave sickness" in 1978 [2]. The International Agency for Research on Cancer, classified RFW as a "possible human carcinogen" which is published in "Agents Classified by the IARC Monographs", Volumes 1–109 (http://monographs.iarc.fr/ENG/Classification/Classifications Alpha Order.pdf). An increase incidence of thyroid cancer in South Korea and gliomas in Sweden were recently reported, and excessive use of mobile phones was held responsible to be a possible cause [12, 13]. Childhood leukemia in children is another disease due to exposer to RFW. Functions of the central nervous system [14], permeability of the blood brain barrier [15], and melatonin synthesis [16] are also affected.

To further study the effects of radio waves, the chick animal model has been used in the past. Laboratories have reported a high mortality of chicken embryos and malformations when exposed to electromagnetic waves emitted from mobile phones. The phenomenon was reported to be dose dependent [17–23]. The embryonic cells rapidly proliferate, differentiate, migrate, and suffer from apoptosis. During these processes, the cells generate electric currents which can be affected by the RFW [24]. This study was designed to observe the effects of electromagnetic field emitted by a mobile phone on rapidly proliferating cells in the developing chicken embryo.

### 1.1 Hypothesis

Rapidly proliferating and dividing cells are affected by RFW exposure, and this may be a dose dependent phenomenon.

## 2. Mobile phone RFW classification

The TriField meter revealed different intensities of electromagnetic waves from different mobile sets. The intensity of RFW was divided into four groups as follows: group 1: 0.01–0.1 $mW/cm^2$, group 2: 0.1–0.2 $mW/cm^2$, group 3: 0.2–0.1 $mW/cm^2$, and group 4: >1 $mW/cm^2$.

All the old mobile sets were placed in group 4 showing the highest levels of radiation (>1 $mW/cm^2$) which is recommended dangerous to health. Mostly, the smart phones were in groups 1 (0.01–0.1 $mW/cm^2$), 2 (0.1–0.2 $mW/cm^2$), and 3 (0.2–1 $mW/cm^2$), but few in group 4. It was further observed that downloading from the net using WiFi also results in high levels of radiations (**Figure 1**).

## 3. Low dose (total 200–300 minutes)

### 3.1 Methods

This is an animal experimental research study using "Cobb" (*Gallus gallus domesticus*) breed zero-day fertilized chicken eggs. These eggs were provided by Sohar Poultry Company S.A.O.G. (PO Box 2808, Ruwi, Postal code 112, Sultanate of Oman). 120 fertilized eggs were used for this experiment. 60 eggs were randomly divided into either the control or the exposed groups.

**Figure 1.**
*TriField meter measuring the electromagnetic field exposure when the phone is receiving a call (the needle moving toward the right with increasing intensity of the electromagnetic waves) (a) group 1, (b) group 2, (c) group 3, (d) group 4, and (e) during downloading, the needle is hitting extreme right.*

A 30-egg incubator Model EH-35, Sino-PFE Company, China was used which is equipped with temperature and humidity control and forced air ventilation; temperature was programed at 37° centigrade and humidity of 50–60% (**Figure 2a**). The experiment was run in batches of 30 eggs, due to the size of the incubator. The eggs were placed in egg holders which were programed to rotate 10 times/day. A mobile phone (power of 0.47 W/kg body and SAR 1.10 W/kg) and local service provider with 1800 MHz frequency were selected. A TriField Meter, model 100XE was used to confirm the emission of electromagnetic field of the mobile phone during the experiment (**Figure 2b**).

### *3.1.1 Exposed group*

The selected mobile phone was placed in the center of the incubator in silent mode and vibration disabled. The farthest egg was within a radius of 16 cm. This

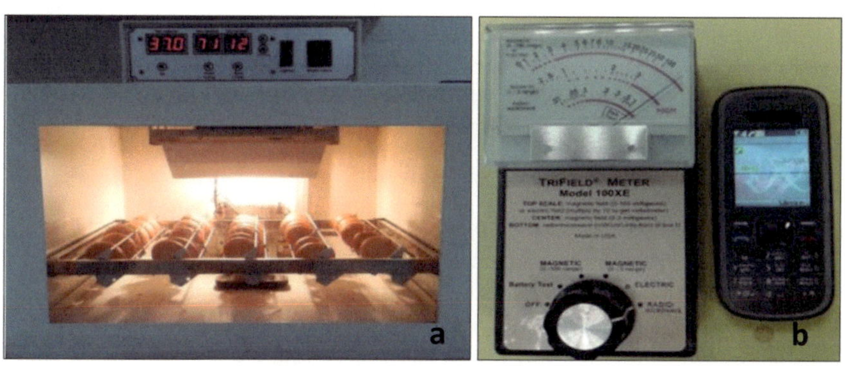


**Figure 2.**
*(a) The egg incubator. (b) TriField meter showing high levels of electromagnetic waves transmitted from the mobile during call receiving (≥1 mW/cm²).*

is important because the radius of single 1800 MHz frequency electromagnetic wavelength is approximately 16.5 cm [21].

The mobile phone placed inside the incubator was called from outside 4 times a day and each call lasted for 5 minutes. There was a gap of 4 hours in-between each call with a call free period during the night. Total exposure time/day was 20 minutes. To repeat the call on the same time every day, a ringing schedule was made, and call was recorded once the call was made. This prevented errors in making the calls. The experiment was conducted in batches of two with 30 eggs in each batch. A total of 10 embryos were scarified each at day 5 (exposure time 100 minutes), day 10 (exposure time 200 minutes), and day 15 (exposure time 300 minutes).

#### *3.1.2 Control group*

It was also conducted in batches of two. There was no mobile placed inside the incubator, and the eggs were not exposed to any electromagnetic field.

A total of 10 randomly selected eggs were removed from the incubator in both the groups at day 5, day 10, and day 15 each, the shells were removed, and the embryos were dissected from the membranes. Mortality of the embryos was observed by observing the movements in the limbs or beating of the heart. Hamburger and Hamilton developmental stages were used to assess the embryos for gross morphological abnormalities [25]. The embryos were washed in normal saline, blotted dry with tissue paper and the dependent variables, the weight, CR-length, and eye diameter were recorded.

A digital balance with a precision of 0.01 g (Universal Impex HA-3202) was used for measuring the weight, and a caliper (Mitutoyo Vernier calipers, Nanjing Sulang Trading Co., Ltd., China) was used to measure the length which was the length from the vertex to the tip of the coccyx and the maximum eye diameter.

For statistical analysis, student's t-test was used, and a p-value of $<0.05$ were considered as significant. All data were presented as the mean value ± SEM.

### 3.2 Results

Mortality was not observed among both the control and the exposed groups, and all 120 eggs revealed developing embryos.

#### *3.2.1 Day 5*

No morphological deformities were observed in both the groups. There was no significant difference between the average wet body weight in the exposed group (0.189 ± 0.035 g) and the control group (0.209 ± 0.031 g); ($t = 1.67$, $df = 28$, $p < 0.11$). The CR-length and eye diameter were not measured at day 5.

#### *3.2.2 Day 10*

Gross malformation was absent in both the groups. In the exposed group, small hemorrhages were apparent under the skin with areas of paler skin indicating reduced blood flow (**Figure 3b**). In the control group, the skin was pink in color throughout (**Figure 3a**). The average weight of the exposed group (1.572 ± 0.38 g) was significantly lower than the weight of the control group (2.331 ± 0.27 g), $t = 8.19$, $df = 48$, $p < 0.01$ (**Figure 5a**). This trend was also observed with the C-R length and eye diameter; all showing significant differences (**Figure 5b** and **c**).

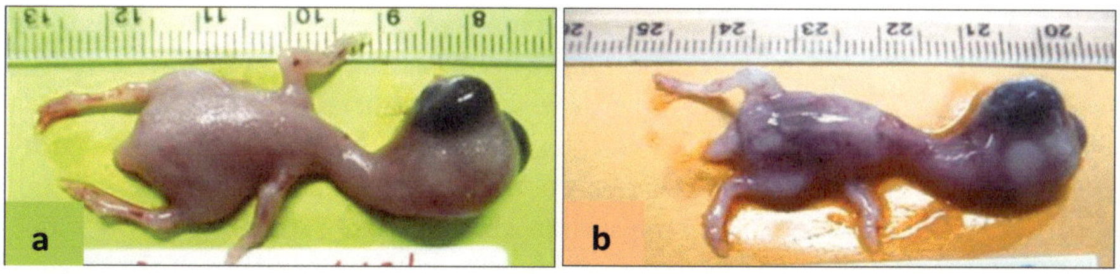


**Figure 3.**
*Low dose: (a) day 10: control group showing normal embryo development. (b) Experimental group: embryos were smaller in size than the control group, and marked hemorrhagic areas could be seen under the skin alternating with pale areas.*

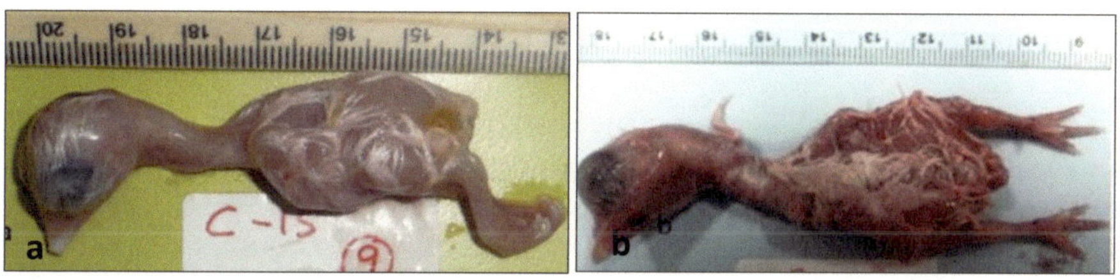


**Figure 4.**
*Low dose: (a) day 15: control group showing well developed embryo with no anomalies or deformities. Skin pink in color, covered by feathers, well-formed upper and lower extremities and normal toes and eyes completely covered by eye lids. (b) Experimental group: no anomalies or deformities were observed and embryo showing same features as revealed in the control group.*

### 3.2.3 Day 15

Gross appearance in both the groups revealed well-formed skin covered by white feathers and well-formed beak. The size of the wings and the limbs was increased, and toes can be seen well formed, middle toe the longest. The head size that was

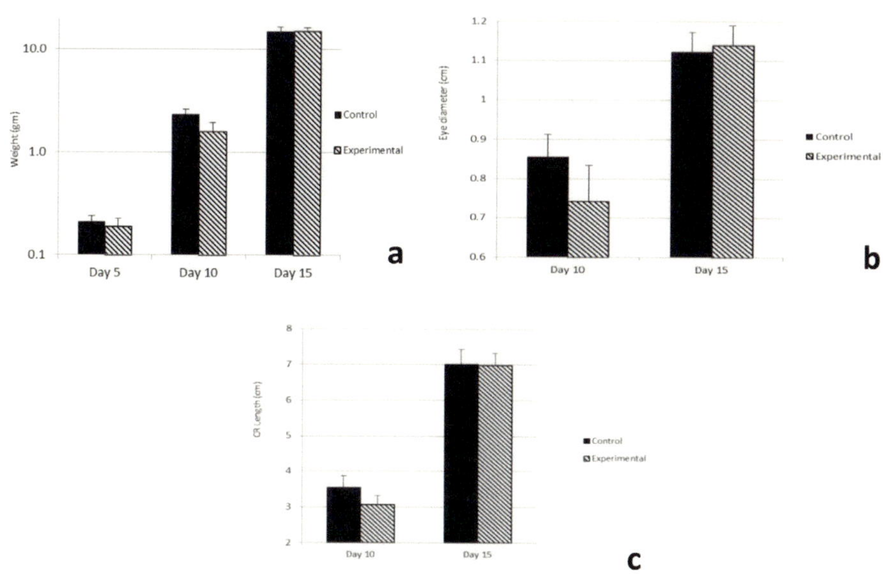


**Figure 5.**
*(a) Wet body weight of the chick embryo: experimental and control groups at days 5, 10, and 15 showing significant different at day 10 ($p < 0.01$). (b) Eye diameter was significantly smaller in the exposed group than the control group at day 10 ($p < 0.01$). (c) C-R length of the chick embryo in the experimental group was significantly smaller than the control group at day 10 ($p < 0.01$).*

much bigger as compared to the body before was now become smaller, which was in accordance with the normal development of the embryo. The eyes were now fully covered by eye lids. The skin no longer shows any hemorrhage areas, and blood vessels no longer visible under the skin (**Figure 4a** and **b**). No significant differences were observed in either average wet body weight, control (14.91 ± 1.73 g) and exposed group (14.82 ± 1.57 g), eye diameter, control (1.123 ± 0.051 cm and exposed 1.14 ± 0.05) and the average C-R lengths, control (7.013 ± 0.41 cm), and the exposed groups (6.978 ± 0.348 cm) (**Figure 5a–c**).

## 4. High dose (total 500–750 minutes)

### 4.1 Methods

A total of 20 fertilized eggs were incubated in the incubator with the mobile phone in silent mode with the vibration mode disabled. The distance of all the eggs from the mobile phone was maintained within one wavelength (approximately 16.5 cm) of the emitting 1800 MHz frequency electromagnetic waves [21]. The mobile phone was rung from another mobile phone for 5 minutes, 10 times daily with an exposure-free period in between the calls. No calls were made at night. The total daily exposure duration was 50 minutes in each 24 hours starting from day 1. The eggs were sacrificed at day 10 (maximum exposure time 500 minutes) and day 15 (total exposure time 750 minutes). For the control groups, another 20 eggs were placed in the incubator, and 10 eggs were sacrificed at day 10 and 15 each.

Control group's 20 eggs were incubated at the same conditions in the same incubator. The mobile phone was turned off, battery removed, and placed in the middle of the incubator. The embryos were examined just as in the experimental groups at days 10 and 15.

On the scheduled day of sacrifice, mortality, gross morphology, wet weight, and length of the embryos were measured. Liver was dissected and placed in 10% glutaraldehyde solution and stained by toluidine blue stain for histological and ultrastructural preparation. Another five specimens were fixed in RNAlater solution (Invitrogen) for heat shock protein 70 (HSP70) and messenger ribonucleic acid (mRNA).

### 4.2 Results

The results revealed 14% mortality in the exposed group. Gross morphology of the chick at day 15 showed deformities of the limbs, hemorrhages under the skin, lack of feathers, and few anterior abdominal wall defects (**Figure 6a** and **b**). The wet weight and head to toe length were significantly less in the exposed group at both day 10 and 15 (**Figure 7a** and **b**).

#### *4.2.1 Histology: control group*

At day 10, hepatocytes were seen with rounded central nucleus and nucleoli. They were lying in rows with spaces in-between to form the sinusoids. Central veins with few RBCs and portal areas were observed. At day 15, typical structure of the liver was apparent. Well-formed hepatic lobules formed by rows of hepatocytes and sinusoids lining with epithelium and large number of RBCs were clearly seen (**Figure 8a** and **b**).

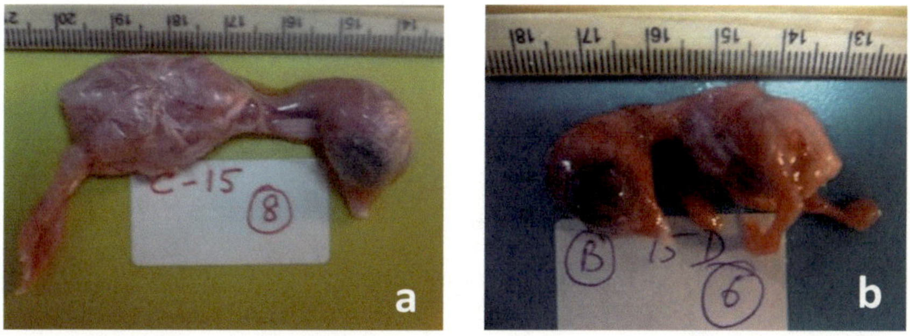


**Figure 6.**
*High dose: (a) day 15: control group showing normal embryo development. (b) Experimental group: embryo was small in size, has marked deformities of the limbs, hemorrhages under the skin, and no growth of the feathers.*

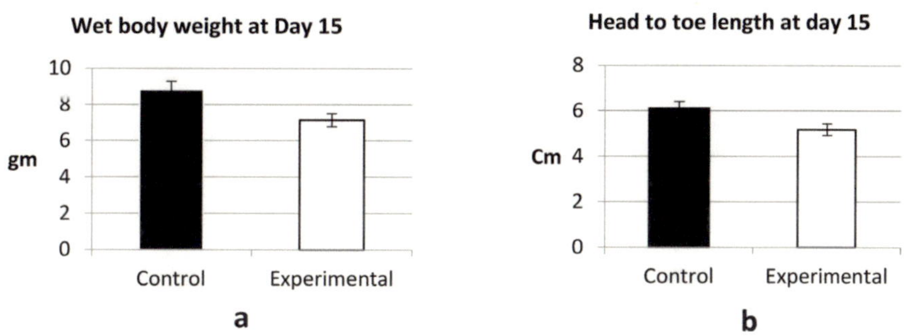


**Figure 7.**
*High dose: (a) wet body weight of the chick embryo: experimental and control groups at days 15 showing significant different (8.84 vs. 7.13 g) (p = 0.29). (b) Head to toe length of the chick embryo in the experimental group was significantly smaller than the control group (6.11 vs. 5.17 cm) at day 15 (p = 0.25).*

### *4.2.2 Exposed group*

At day 10 and day 15, nucleus in many hepatocytes was not seen or pushed to the side and without prominent nucleolus. The hepatocytes were seen in rows with sinusoids in-between; however, marked infiltration of the fat vacuoles was observed in the cytoplasm of hepatocytes. The sinusoids were formed showing lining epithelial cells and RBCs. This signifies the beginning of fatty change (**Figure 8c** and **d**).

### *4.2.3 Electron microscopy: Control group*

At day 10 and 15, the control group showed hepatocytes arranged in rows forming the hepatic lobules and sinusoids in-between the rows lined by simple squamous epithelium (**Figure 9a**). There was a big central rounded nucleus with central chromatin, surrounded by double layered nuclear membrane with pores. Cytoplasm contains many mitochondria with well-arranged cristae, rough endoplasmic reticulum, ribosomes, and some glycogen vacuoles. Sinusoids were formed lined by single layer of simple squamous epithelium, and few Kupffer cells were also found lining the sinusoidal wall. Spindle-shaped RBCs with oval nucleus were present in the sinusoids. The canaliculi were seen clearly in between the hepatocytes (**Figures 9a** and **10a**).

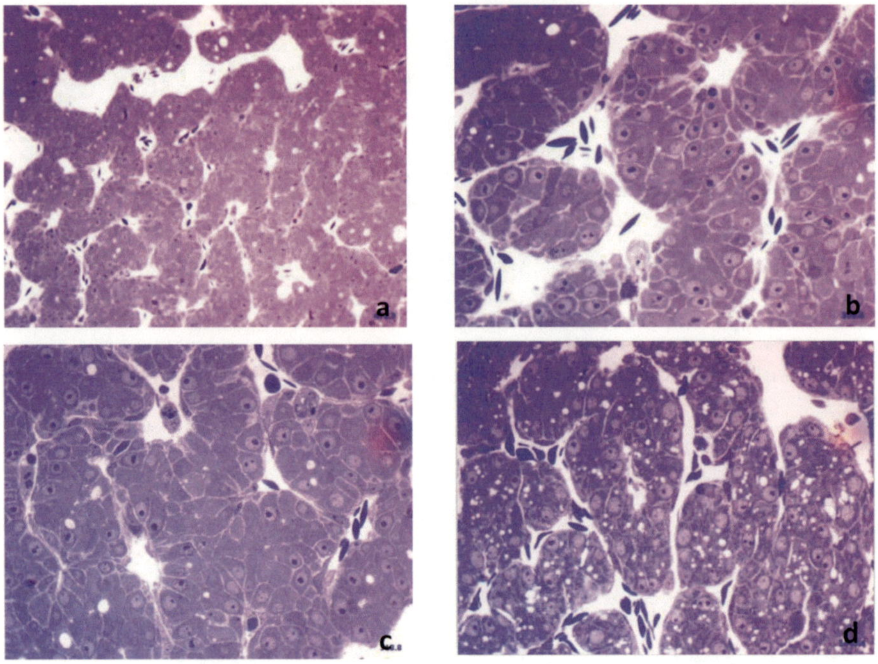


**Figure 8.**
*Control group: (a) day 10 showing developing hepatocytes and sinusoids in between with RBCs and (b) day 15 showing well-formed hepatic lobules, normal hepatocytes arranged in rows, forming central vein, sinusoids lined by epithelial cells, and scattered well-formed RBCs inside the sinusoids. Exposed group: (c) day 10 showing infiltration of few lipid vacuoles in the hepatocytes and few necrotic hepatocytes and (d) day 15 showing marked infiltration of lipids causing necrosis of the hepatocytes.*

### *4.2.4 Exposed group*

At day 10, marked increase in the number of mitochondria can be observed (**Figure 9b**). Some of the mitochondria were swollen and surrounded by rough endoplasmic reticulum. The nucleus was in the center and circular in shape. Sinusoids were filled with red blood cells (RBC) and lined with simple squamous epithelium and Kupffer cells. At day 15, cytoplasm showed numerous lipid filled vacuoles

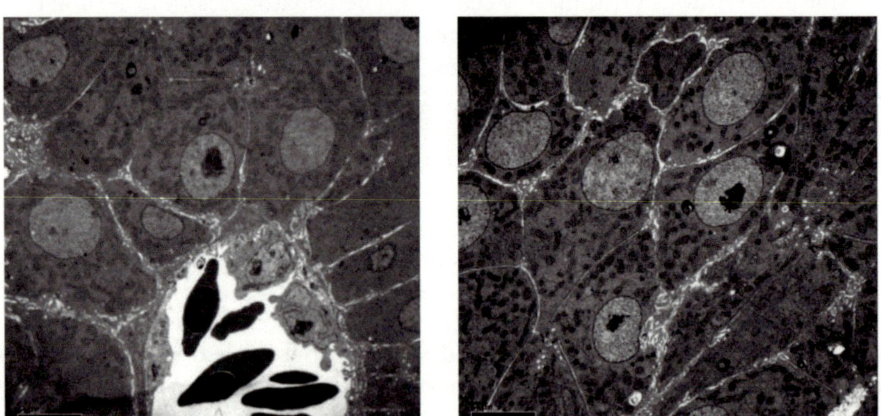

a) b)

**Figure 9.**
*Day 10, (a) in control group, polygonal hepatocytes arranged in row with large central round nucleus and showing sinusoids with multiple red blood cells, and canaliculi can also be seen in-between the hepatocytes. (b) Exposed group revealed marked proliferation of mitochondria and widening of the canaliculi.*

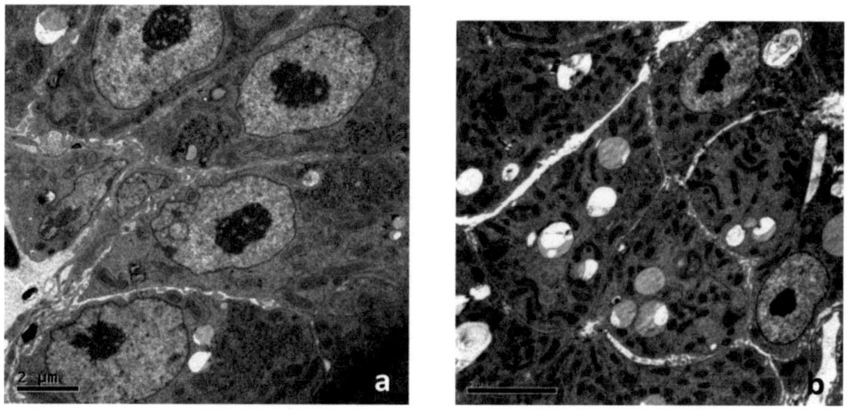


**Figure 10.**
*Day 15, (a) control group showing hepatocytes arranged in rows, with few lipid vacuoles and normal looking canaliculi and (b) exposed group revealed mark infiltration of lipid vacuoles in the cytoplasm, and widening of the canaliculi was apparent.*

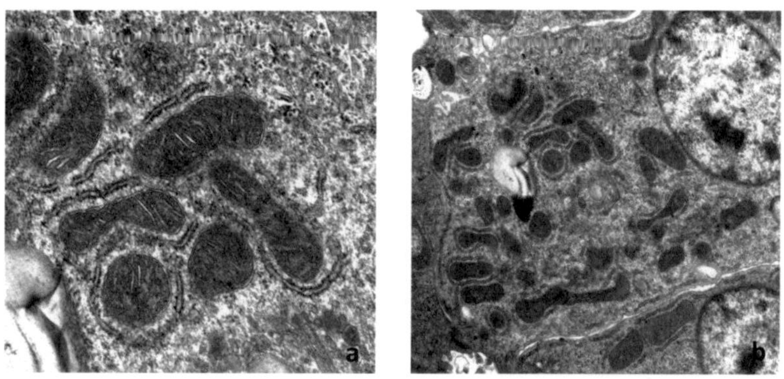


**Figure 11.**
*(a and b) Day 15, exposed group: mitochondria showed change in shape, some become rounded, while others dumbbell shaped and surrounding by rough endoplasmic reticulum. Cristae were not clearly visible, and some areas of degeneration were apparent.*

(**Figure 10b**). At day 15, mitochondria became electron dense, some were swollen, while others were elongated and dumbbell shaped. At places, mitochondria was degenerated and cristae invisible (**Figure 11a** and **b**). A prominent layer of rough endoplasmic layer around the mitochondria could be seen which was interrupted. Myelin-like figures can also be seen in the cytoplasm at day 15 in exposed groups. The canaliculi in-between the hepatocytes have been widened (**Figure 9b**).

## 4.3 Biochemical analysis

### *4.3.1 Heat shock protein 70*

Homogenized sample using an automated homogenizer in a phosphate buffer at pH 7.6 was ultracentrifugated to separate contaminants and soluble proteins and cells. Messenger RNA (mRNA) was extracted from 40 tissue samples using the Qiagen RNeasy Mini Kit (QIAGEN, CA, USA) according to the manufacturer's instructions.

Next, DNAse treatment was given to the extracted mRNA using DNA-free™ DNA Removal Kit (Thermo fisher) according to the manufacturer's protocol. After which, cDNA was synthesized using *DNase treated RNA with the help of* High-Capacity cDNA Reverse Transcription Kit (Applied BioSystems, Austin, TX).

*Quantitative RT-PCR was performed for the analysis of cDNA obtained from the 40 samples.* TaqMan reagents are used to perform qRT-PCR according to manufacturer's protocol. The relative expression level of *hsp70* was calculated using the comparative delta Ct method by normalizing the cycle threshold values of *hsp70* with those of GAPDH.

Protein extraction was carried out through differential centrifugation to obtain separate subcellular fractions of protein from hepatocytes.

Extracted proteins were analyzed through SDS-PAGE and one step ELISA. HSP70—ab187399 Simple Step ELISA® Kit was used for the detection of HSP70 protein in the control and experimental samples.

*4.3.2 Results*

Different exposure durations displayed significantly different levels of HSP 70 mRNA compared to the normal. This increase in mRNA was 14% at day 10 and became 39% at day 15 (**Figure 12**). The amount of HSP70 protein has increased with longer exposure time although it did not prove strong correlation with the

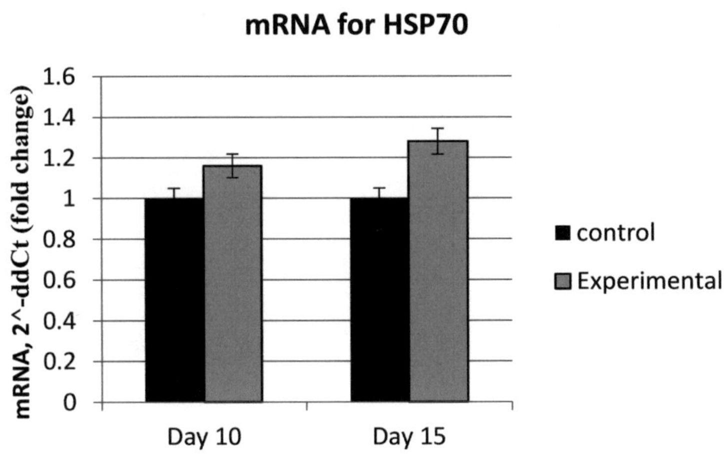


**Figure 12.**
*mRNA for HSP70 for liver tissue at day 10 and day 15 in the control and experimental groups.*

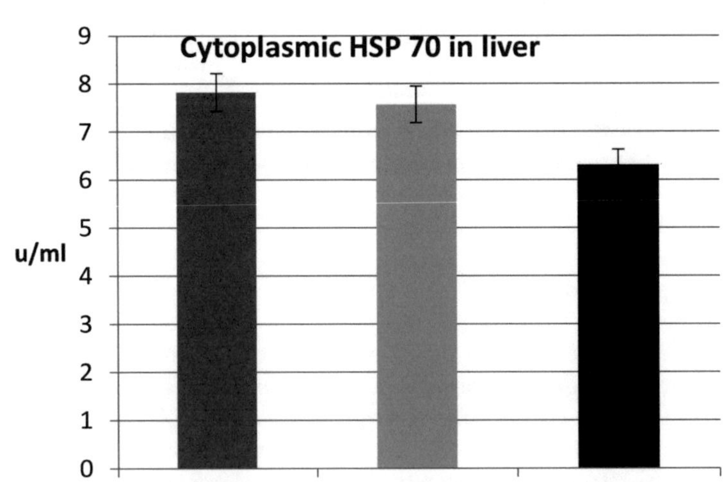


**Figure 13.**
*ELISA graphs showing increase in the cytosolic HSP70 in the control and exposed groups in liver at day 10 and day 15.*

expressed mRNA quantities (**Figure 13**). The total cellular protein represented by the cytoplasmic protein concentration was increased in treated samples than in untreated samples demonstrating cellular mechanism in attempting to express housekeeping genes and HSP70 for survival. Organs have shown differences. Our results indicate that HSP 70 mRNA is over expressed in an attempt to synthesize more of the HSP70 in order to resist electromagnetic waves irradiation.

## 5. Discussion

### 5.1 Measuring mobile phones' RFW strength

To confirm the emission of RFW from different mobile sets and to understand its strength, we decided to measure the RFW from different mobile sets using a TriField meter. It was found that the old mobile sets were emitting very high intensity of RFW which the TriField meter showing above 1 $mW/cm^2$. Absolute hazard thresholds have not been established yet; however, studies suggest that RFW above 0.1 $mW/cm^2$ may not be safe. Our study revealed that most of the smart mobile sets are emitting RFW above 0.1 $mW/cm^2$. According to our classification, only group 1 comes under 0.1 $mW/cm^2$; groups 2, 3 and 4 are all above this threshold.

### 5.2 Mortality

In the low dose groups (exposure time 20 minutes/day at 1800 MHz), no mortality was observed in either the control or experimental groups. Batellier et al. also reported a mortality rate of less than 1% when fertilized eggs were exposed to 900 MHz from day 7 to 14 [13].

In the higher dose group, the mortality rose to 14% in the exposed group, while no mortality was present in the control group. About 54% mortality was reported by Youbicier-Simo in the exposed group and 14% in the control group. He used continuous mobile phone exposure during the embryonic life of 21 days [14]. Higher mortality was also reported by Jyoti et al. in the exposed group. He highlighted that increased exposure duration and higher power (20 dBm) must have increased mortality in the exposed group [15].

### 5.3 Gross malformations

In the low dose group, the most significant finding in this study was growth retardation of the embryo at day 10 only; body weight, CR length, and eye diameter were all significantly decreased at day 10 in the exposure group. However, they were indistinguishable from the control group at both day 5 and day 15. No gross malformation was observed except subcutaneous hemorrhagic areas under the skin at day 10.

Cox et al. reported no malformations when he exposed the fertilized eggs to 50 Hz [16]. Al Qusdi et al. using 900–1800 MHz electromagnetic waves and by ringing 4 times for 15 minutes/day (60 minutes/day) reported significant increase in body weight and length at day 10, which could not be sustained at day 14 [17]. Gross malformation in the brain and retina was also reported; however, the incubator used in that study was not an egg incubator, which could have negative effects on the development of the embryo [17].

In high dose group in this study, 14% mortality, limb deformities, and changes in the skin feathers were observed. There was growth retardation of the embryo at both day 10 and day 15. Other studies also reported growth retardation [18]. Other malformations reported by Lahijani et al. such as spina bifida, monophthalmia, microphthalmia, anophthalmia, exencephalic embryos, and embryos with asymmetrical faces, crossed or shorter beak, and gastroschisis were not found in this study [18]. In an extensive study done by Farrell et al. at five different laboratories and using 60 Hz magnetic field (pulsed and sinusoidal by a Tenma function generator) over 2500 chick embryos, did not show significant difference in the mortality for 48 hours; however, over the period of time, malformations were 6.8% in the exposed group compared to 1.8% in the control; most common anomaly reported was the neural tube defects [19].

Ubeda et al. observed in a study using 100 Hz and electromagnetic field intensity between 0.4 and 104 microTeslas (μT) that the chick embryo is sensitive to electromagnetic fields at extremely low frequency and intensity. He further stated that pulse shape may influence the development of embryo development [20].

Effects of electromagnetic waves on living cells are dose and duration dependent [21]. It is likely that growth retardation of the chick embryo in exposed groups was a result of interference in the proliferation of the multiplying embryonic cells due to the RFW exposure. Development of the embryo is a complicated process which includes cell proliferation, differentiation, relocation, and programed cell death. These events involve endogenous ionic currents and electric fields which could be disturbed by the RFW exposure. Growth retardation in the exposed group is most likely due to the adverse effects of RFW on the DNA. Production of heat shock proteins is stimulated to repair this damage. This also increases reactive oxygen species production and apoptosis [17, 22, 23]. DNA damage if not repaired would most likely result in cell death [23]. This cell self-repair depends on the intensity of the initial injury by the radio waves. Thus at day 10, the chick embryo cells proliferation and multiplication declined due to DNA damage by the RFW exposure, hence reducing the wet body weight of the embryo as compared to the control group. At day 15, the wet body weight, CR length, and eye diameter did not show significant difference from the control in low dose because the cells were able to repair the damage; however, in the high dose group, the damage could not be repaired and significant difference persisted at day 15.

## 5.4 Histology and electron microscopy

In this study, an increase number of mitochondria were observed in the exposed group at day 10 and day 15. It has been reported that oxidative stress induced by $H_2O_2$ leads to increase number of mitochondria and mtDNA in human lung [24]. At day 10 in the exposed group, some of the mitochondria were swollen. It has been reported that permeability of mitochondrial membrane is dependent on interaction between $Ca^{+2}$ and reactive oxygen species (ROS) system. Increased sympathetic activity is considered a primary cause of REW-induced calcium influx into the mitochondria [26]. ROS when stimulated on inner surface of mitochondria will produce free $O^{+2}$ which invade thiol protein and cause transition pores in the membrane to open thus increasing its permeability and causing mitochondria swelling [26]. Mitochondrial swelling was also observed by Atti [27] in rat hepatocytes in response to electromagnetic field exposure.

At day 15, in the exposed group, most of the mitochondria were elongated, dumbbell shaped, and in the process of degeneration. They were surrounded by fragments of rough endoplasmic reticulum and lots of free ribosomes. It was reported that mitochondria is the source of free radicals produced in response to electromagnetic wave exposure in human sperms [22]. It further reported potential causative mechanism of electron leakage from the mitochondrial electron transport chain which causes oxidative DNA damage.

The next most significant alteration in the hepatocytes was the increase number of lipid filled vacuoles in the exposed groups. It was reported that this marked increase in the cytoplasmic vacuolation is due to disturbances in lipid inclusions and fat metabolisms in pathological conditions, blockage of gluconeogenesis due to free radicals changing the nature of lipids, protein groups, and cell damage [26]. It seems that rough endoplasmic reticulum surrounding the mitochondria is damaged and fragmented thus contributing in increase in fat droplets.

Electromagnetic waves caused fatty change in the hepatocytes of the developing chick embryos. In fatty liver, there is an increase in lipid droplets in the cytoplasm of the hepatocytes suggesting that cells are under oxidative stress when exposed to electromagnetic waves [28–30]. Lahijani et al. had similar results showing abnormal lipid droplets in the hepatocyte cytoplasm and pushing the nuclei to one side [26]. Similar results were reported in rats and rabbits [22, 31]. The breakdown of fat in the liver may be disrupted by radiation exposure, which may be similar to alcoholism, malnutrition, poisoning, and pregnancy. Fatty change is the beginning of injury to the hepatocytes, showing an increase in vacuoles filled with triglyceride fat, a sign of abnormal metabolism which may be due to the production of oxygen radicles species in the hepatocytes [29–31].

### 5.5 Biochemical test

Heat shock protein70 (HSP70) is a multifaceted chaperon protein that is conserved across species and is involved in many cellular metabolic processes. Its prime role as an anti-stress agent has been exploited heavily in an attempt to further clarify its mode of action and cellular distribution [33]. In this study, Hsp70 was also investigated for its role and degree of expression in stress conditions induced by electromagnetic waves. Previous studies have demonstrated the overexpression of Hsp70 mRNA under stress stimuli but have not elucidated the magnitude and the subcellular compartment localization of the protein [34, 35].

In this study, we observed significant increase in levels of mRNA and Hsp70 in the exposed groups at both day 10 and 15. This increase in mRNA was 14% at day 10 and became 39% at day 15 (**Figure 13**). Levels of mRNA and HSP70 in the liver tissue are demonstrating an important rule liver plays in combating cellular stress using HSP70 as one of the approaches to attenuate the stress stimuli (**Figure 12**).

Mobile phone radiation induces reactive oxygen species (ORS) and DNA damage as seen by increase in the HSP70 and its mRNA and which causes metabolic, immunological, and carcinogenic effects [2, 36–38]. ORS was also reported in human sperm, which affects genes, cell membrane function, and signal transduction [28, 39, 40]. Rao et al. recently provided new evidence supporting the theory that radio waves affect the plasma membrane [39]. Radio waves also induce NADH oxidase enzyme, which might play a key role in the various cellular adverse effects [36, 40]. Various cellular and physiological processes can be affected as a consequence of increased levels of free radicals, including gene expression, cell growth, apoptosis, and release of calcium from intracellular storage sites [32, 36, 41–49].

## 6. Conclusion

We conclude that even at a low dose of radio waves (20 minutes/day) emitted by mobile phones, the development of chick embryos as seen on the 10th day of incubation would be affected. However, the cells were able to repair the damage by day 15 and no significant difference was observed. At higher dose, the significant different persisted at day 15, and mortality was increased with morphological changes observed. Histological finding in the liver revealed fatty change which was confirmed by ultrastructural findings. The increase of HSP 70 and mRNA in the exposed groups supported the assumption that RFW exposure has increased ORS which caused damage to the organelles and metabolism in the hepatocytes. These effects question the safety of mobile phones and their potential as a hazard to living cells. It is recommended that mobile phones should be used with caution.

## Acknowledgements

I am grateful to Dr. Mubarak Pasha, Deputy Dean Administration for providing me with the laboratory equipment and tools to conduct this research. I am thankful to Dr. Mark Norrish, who at present is Associate Dean International at Coventry University, UK worked as Associate Professor at Oman Medical College in Oman, for his expertise in doing statistical analysis and Dr. John Mutusami, Head Department of Surgery, for his critical review and analysis. Special thanks to Dr. Azem Shalaby, Associate Professor at Department of Pathology, Dr. Irfan Ullah, Department of Pediatrics, Dr. Yahya Tamimi, Associate Professor of Biochemistry and College of Medicine and Health Sciences, Sultan Qaboos University for their strong collaboration in this research project. I am also in debt to our Dean, Dr. Salah Al Kusaiby, and Vice-Dean, Mohammad Al Shafaee, Department of Family Medicine, and Prof. ML Ajmani, Head of Department of Anatomy and Neurobiology, College of Medicine and Health Sciences, National University of Health Sciences, Oman, for their encouragement and endless support throughout this study. In the end I sincerely thank Mr. Mohammad Al Kindi, Ms. Fatima Al-Ghafri, Mr. Abdulrahman Al-Nabhani, Electron Microscopy Laboratory, Department of Pathology, and Ms. Zoya Sheikh, Biochemistry Department, Sultan Qaboos University, without their help I could never have completed this study. I am also thankful to all my MD students who chipped in from time to time and ensured continuous progress of this research. The first student who started this research was Dr. Syed M Saud, who also received 2nd prize in oral presentation at 8th GCC Medical Student's Conference at SQU, Muscat, Oman. Other students who participated were Dr. Ayesha Shafaq, Dr. Marwan Zaki, Dr. Afaf Al Farsi, Ms. Fatima Al Baloshi, Ms. Sakoon Said Alharthi, Mr. Mohammed Amir Rafe, Mr. Muhammad Hamza Kamal, Ms. Rana Hamood Al Maqbali, Ms. Reem Hamood Al Maqbali, and Ms. Sara Al Adawai.

## Conflict of interest

There is no conflict of interest. This research was approved by the Ethical Research Board of the Medical College and supported by an internal grant.

## Author details

Najam Siddiqi[1*] and Nasser Al Nazwani[2]

1 Department of Anatomy and Neurobiology, National University of Science and Technology, Sohar, Oman

2 Department of Biochemistry, National University of Science and Technology, Sohar, Oman

*Address all correspondence to: najamsiddiqi@nu.edu.om

## References


[1] Samkange-Zeeb F, Blettner M. Emerging aspects of mobile phone use. Emerging Health Threats Journal. 2009;**2**:2-8

[2] Blake LB, Lai H. Biological effects from exposure to electromagnetic radiation emitted by cell tower base stations and other antenna arrays. Environmental Reviews. 2010;**18**:369-397

[3] Leitgeb N. Mobile phones: Are children at higher risk? Wiener Medizinische Wochenschrift. 2008;**158**(1-2):36-41

[4] Kheifets L, Repacholi M, Saunders R, Van deventer E. The sensitivity of children to electromagnetic fields. Pediatrics. 2005;**116**:303-313

[5] Maisch D. Children and mobile phones... is there a health risks? Journal of Australian College of Nutritional and Environmental Medicine. 2003;**22**:3-8

[6] Siddiqi N, Heming T. Effects of mobile phone radiofrequency waves (REW) on embryonic stem cells. The FASEB Journal. 2013;**27**:753.1

[7] Schüz J. Mobile phone use and exposures in children. Bioelectromagnetics Supplement. 2005;**7**:S45-S50

[8] Abramson MJ, Benke GP, Dimitriadis C, et al. Mobile telephone use in associated with changes in cognitive function in young adolescents. Bioelectromagnetics. 2009;**30**:678-686

[9] Divan HA, Kheifets L, Obel C, Olsen J. Prenatal and postnatal exposure to cell phone use and behavioral problems in children. Epidemiology. 2008;**19**(4):523-529

[10] World Health Statistics. World Health Organization, WHO Library Cataloguing-in-Publication Data; 2013

[11] Ngole I, Ghosh SK. Exposure to radio frequency radiation emitted by cell phone and mortality in chick embryo (*Gallus domesticus*). Biomedical Research. 2006;**17**:205-210

[12] Oh CM, Jung KW, Won YJ, Shin A, Kong HJ, Lee JS. Age-period-cohort analysis of thyroid cancer incidence in Korea. Cancer Research and Treatment. 2015;**47**(3):362-369

[13] Hardell L, cariberg M. Mobile phone and cordless phone use and the risk for glioma–Analysis of pooled case-control studies in Sweden, 1997-2003 and 2007-2009. Pathophysiology. 2015;**22**(1):1-13

[14] Maier R. Is CNS activity modified by pulsed electromagnetic fields? Biomedizinische Technik. Biomedical Engineering. 2001;**46**(1-2):18-23

[15] Franke H, Streckert J, Bitz A, Goeke J, Hansen V, Ringelstein EB, et al. Effects of universal mobile telecommunications system (UMTS) electromagnetic fields on the blood-brain barrier in vitro. Radiation Research. 2005;**164**(3):258-269

[16] Sukhotina I, Streckert JR, Bitz AK, Hansen VW, Lerchl A. A 1800 MHz electromagnetic field effects on melatonin release from isolated pineal glands. Journal of Pineal Research. 2006;**40**(1):86-91

[17] Zareen N, Khan MY, Minhas LA. Dose related shifts in the developmental progress of chick embryo exposed to mobile phone induced electromagnetic fields. Journal of Ayub Medical College. 2009;**21**(1):130-134

[18] Hamburger V, Hamilton HL. A series of normal stages in the development of the chick embryo. Developmental Dynamics. 1992;**195**(4):231-272

[19] Batellier F, Couty I, Picard D, Brillard JP. Effects of exposing chicken

eggs to a cell phone in "call" position over the entire incubation period. Theriogenology. 2008;**69**(6):737-745

[20] Youbicier-Simo BJ, Lebecq JC, Giaimis J, et al. Mortality of chicken embryos continuously exposed under GSM cell phone and validation of the effectiveness of a protective device. In: Proceeding of International Conference on Cell Tower Siting; Salzburg, Austria. 2000. pp. 233-235

[21] Jyoti KR, Bagai U. Effect of mobile phone frequency radiation on early development of chick embryo. International Journal of Science, Environment and Technology. 2014;**3**:1273-1280

[22] Cox CF, Brewer LJ, Raeman CH, Schryver CA, Child SZ. A test for eratology effects of power frequency magnetic fields on chick embryos. IEEE Transactions on Biomedical Engineering. 1993;**40**(7):605-610

[23] Al-Qudsi F, Azzouz S. Effects of electromagnetic radiation on chick embryo development. Life Sciences. 2012;**9**:983-991

[24] Lahijani MS, Farivar S, Khodaeian M. Effects of 50 Hz electromagnetic fields on the histology, apoptosis, and expression of c-Fos and β-catenin on the livers of preincubated white Leghorn chicken embryos. Electromagnetic Biology and Medicine. 2011;**30**(3):158-169

[25] Farrell JM, Litovitz TL, Penafiel M, Montrose CJ, Doinov P, Barber M, et al. The effect of pulsed and sinusoidal magnetic fields on the morphology of developing chick embryos. Bioelectromagnetics. 1997;**18**(6):431-438

[26] Ubeda A, Leal J, Trillo MA, et al. Pulse shape of magnetic fields influences chick embryogenesis. Journal of Anatomy. 1983;**137**:513-536

[27] De Iuliis GN, Newey RJ, King BV, Aitken RJ. Mobile phone radiation induces reactive oxygen species production and DNA damage in human spermatozoa in vitro. PLoS One. 2009;**4**(7):e6446

[28] Aitken RJ, Bennetts LE, Sawyer D, Wiklendt AM, King BV. Impact of radio frequency electromagnetic radiation on DNA integrity in the male germline. International Journal of Andrology. 2005;**28**(3):171-179

[29] Lee H-C, Yin P-H, Lu C-Y, Chi C-W, Wei Y-H. Increase of mitochondria and mitochondrial DNA in response to oxidative stress in human cells. The Biochemical Journal. 2000;**348**:425-432

[30] Lahijani MS, Tehrani DM, Sabouri E. Histopathology and ultrastructural studies on the effects of electromagnetic fields on the liver of pre-incubated white Leghorn chicken embryo. Electromagnetic Biology and Medicine. 2009;**28**:391-413

[31] Atti AA, Yehia MA. Histological, ultrastructural and immune-histochemical studies of the low frequency electromagnetic field effect on thymus, spleen and liver of albino swiss mice. Pakistan Journal of Biological Sciences. 2002;**5**(9):931-937

[32] De luliis GN, Newey RJ, King BV, John Aitken R. Mobile phone radiation induces reactive oxygen species production and DNA damage in human spermatozoa in vitro. PLoS One. 2009;**4**(7):e6446

[33] Angulo P. Nonalcoholic fatty liver disease. The New England Journal of Medicine. 2002;**346**:1221-1231

[34] Reddy JK, Rao MS. Lipid metabolism and liver inflammation. II. Fatty liver disease and fatty acid oxidation. American Journal of Physiology. Gastrointestinal and Liver Physiology. 2006;**290**:852-858

[35] Ibrahim M, El-Ashry M, Ali E. The influence of 50 Hz magnetic field on liver function. Romanian Journal of Biophysics. 2008;**18**:113-122

[36] Tarantino P, Lanubile R, Abbro L, Dini L. Post-continuous whole body exposure of rabbits to 60 MHz electromagnetic fields: Effects on liver, spleen and brain. Radiation and Environmental Biophysics. 2005;**44**:51-59

[37] Brenu EW, Staines DR, Tajouri L, et al. Heat shock proteins and regulatory T cells. Autoimmune Disease. 2013;**813**:256

[38] Lantow M, Schuderer J, Hartwig C, Simkó M. Free radical release and HSP70 expression in two human immune-relevant cell lines after exposure to 1800 MHz radiofrequency radiation. Radiation Research. 2006;**165**(1):88-94

[39] Finka A, Mattoo RUH, Goloubinoff P. Meta-analysis of heat- and chemically upregulated chaperone genes in plant and human cells. Cell Stress & Chaperones. 2011;**16**:15-31

[40] Kiang JG, Gist ID, Tsokos GC. Regulation of heat shock protein 72 kDa and 90 kDa in human breast cancer MDA-MB-231 cells. Molecular and Cellular Biochemistry. 2000;**204**:169-178

[41] Spadaro JA, Bergstrom WH. In vivo and in vitro effects of a pulsed electromagnetic field on net calcium flux in rat calvarial bone. Calcified Tissue International. 2002;**70**(6):496-502

[42] Bonhomme-Faivre L, Marion S, Forestier F, Santini R, Auclair H. Effects of electromagnetic fields on the immune systems of occupationally exposed humans and mice. Archives of Environmental Health. 2003;**58**(11):712-717

[43] Shangcheng X, Zhou Z, Lei Z, Zhengping Y, et al. Exposure to 1800 MHz radiofrequency radiation induces oxidative damage to mitochondrial DNA in primary cultured neurons. Brain Research. 2010;**1311**:189

[44] Rao VS, Titushkin IA, Moros EG, Pickard WF, Thatte HS, Cho MR. Non-thermal effects of radiofrequency-field exposure on calcium dynamics in stem cell-derived neuronal cells: Elucidation of calcium pathways. Radiation Research. 2008;**169**(3):319-329

[45] Dasdag S, Akdag MZ. The link between radiofrequencies emitted from wireless technologies and oxidative stress. Journal of Chemical Neuroanatomy. 2016;**75**:85

[46] Verschaeve L, Heikkinen P, Verheyen G, Van Gorp U, et al. Investigation of co-genotoxic effects of radiofrequency electromagnetic fields in vivo. Radiation Research. 2006;**165**:598-607

[47] Donato A, Ceci P, Cannavo A, Tomei F, Naro F. Low power microwave interaction with phospholipase C and D signal transduction pathways in myogenic cells. Cell Biology International. 2004;**28**:683-688

[48] Markova E, Malmgren LO, Belyaev IY. Microwaves from mobile phones inhibit 53BP1 focus formation in human stem cells more strongly than in differentiated cells: Possible mechanistic link to cancer. Environmental Health Perspectives. 2010;**118**(3):394-399

[49] Desai N, Sharma R, Makker K, Sabanegh E, Agarwal A. Physiological and pathological levels of reactive oxygen species in neat semen of infertile men. Fertility and Sterility. 2009;**92**(5):1626-1631